Testing R Code

Chapman & Hall/CRC
The R Series

Series Editors

John M. Chambers
Department of Statistics
Stanford University
Stanford, California, USA

Torsten Hothorn
Division of Biostatistics
University of Zurich
Switzerland

Duncan Temple Lang
Department of Statistics
University of California, Davis
Davis, California, USA

Hadley Wickham
RStudio
Boston, Massachusetts, USA

Aims and Scope

This book series reflects the recent rapid growth in the development and applicatio of R, the programming language and software environment for statistical computin and graphics. R is now widely used in academic research, education, and industry It is constantly growing, with new versions of the core software released regularl and more than 7,000 packages available. It is difficult for the documentation t keep pace with the expansion of the software, and this vital book series provides forum for the publication of books covering many aspects of the development an application of R.

The scope of the series is wide, covering three main threads:
- Applications of R to specific disciplines such as biology, epidemiology, genetics, engineering, finance, and the social sciences.
- Using R for the study of topics of statistical methodology, such as linear and mixed modeling, time series, Bayesian methods, and missing data.
- The development of R, including programming, building packages, and graphics.

The books will appeal to programmers and developers of R software, as well a applied statisticians and data analysts in many fields. The books will featur detailed worked examples and R code fully integrated into the text, ensuring the usefulness to researchers, practitioners and students.

Published Titles

Stated Preference Methods Using R, *Hideo Aizaki, Tomoaki Nakatani, and Kazuo Sato*

Using R for Numerical Analysis in Science and Engineering, *Victor A. Bloomfield*

Event History Analysis with R, *Göran Broström*

Extending R, *John M. Chambers*

Computational Actuarial Science with R, *Arthur Charpentier*

Testing R Code, *Richard Cotton*

Statistical Computing in C++ and R, *Randall L. Eubank and Ana Kupresanin*

Basics of Matrix Algebra for Statistics with R, *Nick Fieller*

Reproducible Research with R and RStudio, Second Edition, *Christopher Gandrud*

R and MATLAB® *David E. Hiebeler*

Statistics in Toxicology Using R *Ludwig A. Hothorn*

Nonparametric Statistical Methods Using R, *John Kloke and Joseph McKean*

Displaying Time Series, Spatial, and Space-Time Data with R, *Oscar Perpiñán Lamigueiro*

Programming Graphical User Interfaces with R, *Michael F. Lawrence and John Verzani*

Analyzing Sensory Data with R, *Sébastien Lê and Theirry Worch*

Parallel Computing for Data Science: With Examples in R, C++ and CUDA, *Norman Matloff*

Analyzing Baseball Data with R, *Max Marchi and Jim Albert*

Growth Curve Analysis and Visualization Using R, *Daniel Mirman*

R Graphics, Second Edition, *Paul Murrell*

Introductory Fisheries Analyses with R, *Derek H. Ogle*

Data Science in R: A Case Studies Approach to Computational Reasoning and Problem Solving, *Deborah Nolan and Duncan Temple Lang*

Multiple Factor Analysis by Example Using R, *Jérôme Pagès*

Customer and Business Analytics: Applied Data Mining for Business Decision Making Using R, *Daniel S. Putler and Robert E. Krider*

Implementing Reproducible Research, *Victoria Stodden, Friedrich Leisch, and Roger D. Peng*

Graphical Data Analysis with R, *Antony Unwin*

Using R for Introductory Statistics, Second Edition, *John Verzani*

Advanced R, *Hadley Wickham*

Dynamic Documents with R and knitr, Second Edition, *Yihui Xie*

Testing R Code

Richard Cotton
DataCamp
Cambridge, Massachusetts, USA

CRC Press
Taylor & Francis Group
Boca Raton London New York

CRC Press is an imprint of the
Taylor & Francis Group, an **informa** business

A CHAPMAN & HALL BOOK

CRC Press
Taylor & Francis Group
6000 Broken Sound Parkway NW, Suite 300
Boca Raton, FL 33487-2742

© 2017 by Taylor & Francis Group, LLC
CRC Press is an imprint of Taylor & Francis Group, an Informa business

No claim to original U.S. Government works

Printed on acid-free paper
Version Date: 20161202

International Standard Book Number-13: 978-1-4987-6365-3 (Hardback)

Visit the Taylor & Francis Web site at
http://www.taylorandfrancis.com

and the CRC Press Web site at
http://www.crcpress.com

Printed and bound in Great Britain by
TJ International Ltd, Padstow, Cornwall

Contents

Preface   **xiii**

1 Introduction   **1**
  1.1 Interacting vs. Programming   1
  1.2 Two Kinds of Testing   2
  1.3 How Run-Time Testing Will Help You   2
  1.4 How Development-Time Testing Will Help You   4
  1.5 Summary   7

2 Run-Time Testing with *assertive*   **9**
  2.1 Using Assertions   9
     2.1.1 Chaining Assertions Together with Pipes   10
  2.2 Using Predicates: *is* and *has* Functions   11
     2.2.1 Exercise: Using Predicates and Assertions   13
  2.3 The Virtual Package System   14
  2.4 A Tour of the *assertive* Package   14
     2.4.1 assertive.base   14
     2.4.2 assertive.properties   16
     2.4.3 assertive.types   19
     2.4.4 assertive.numbers   20
     2.4.5 Exercise: Examining an Object   22
     2.4.6 assertive.files   22
     2.4.7 Exercise: Inspecting Files and Directories   23
     2.4.8 assertive.strings   24
     2.4.9 assertive.matrices   25
     2.4.10 Exercise: Testing Properties of Matrices   25
     2.4.11 assertive.sets   25
     2.4.12 assertive.models   26
     2.4.13 assertive.reflection   27
     2.4.14 Exercise: Explore Your R Setup   28
     2.4.15 assertive.datetimes   28
     2.4.16 assertive.data, assertive.data.us, and assertive.data.uk   29
     2.4.17 Exercise: Checking Customer Data   30
     2.4.18 assertive.code   31
  2.5 Controlling Severity   32
  2.6 Fail Early, Fail Often   33

2.7 Case Study: Calculating the Geometric Mean 34
    2.7.1 Exercise: Calculating the Harmonic Mean 37
2.8 Alternatives 37
2.9 Summary 38

3 Development-Time Testing with testthat   39
3.1 Using Unit Tests 39
3.2 The Structure of a Unit Test 39
    3.2.1 Exercise: Using `expect_equal` 40
    3.2.2 How Equal Is Equal? 41
3.3 Testing Errors 42
    3.3.1 Exercise: Using `expect_error` 42
3.4 Different Types of Expectation 42
3.5 Testing Warnings and Messages 43
    3.5.1 Chaining Expectations Together with Pipes 44
    3.5.2 Exercise: Using `expect_output` 45
    3.5.3 Testing for Lack of Output 45
3.6 Varying the Strictness of Expectations 46
3.7 Providing Additional Information on Failure 47
3.8 Case Study: Calculating Square Roots 48
    3.8.1 Exercise: Find More Tests for `square_root` 53
3.9 Other Expectations 54
    3.9.1 Exercise: Testing the Return Type of Replicates . . . 56
3.10 Organising Tests Using Contexts 57
3.11 Running Your Tests 57
3.12 Customizing How Test Results Are Reported 59
3.13 Alternatives 60
3.14 Summary 61

4 Writing Easily Maintainable and Testable Code   63
4.1 Don't Repeat Yourself 63
    4.1.1 Case Study: Drawing Lots of Plots 64
    4.1.2 Idea 1: Use Variables Rather Than Hard-Coded Values 68
    4.1.3 Idea 2: For Values That You Want To Change Every-
           where, Update Global Settings 69
    4.1.4 Idea 3: Wrap the Contents into a Function 70
    4.1.5 Exercise: Reducing Duplication 71
4.2 Keep It Simple, Stupid 72
    4.2.1 Simplifying Function Interfaces 72
    4.2.2 Idea 1: Pass Arguments for Advanced Functionality to
           Another Function 74
    4.2.3 Exercise: Outsourcing Argument Checking 76
    4.2.4 Idea 2: Having Wrapper Functions for Specific Use
           Cases 76
    4.2.5 Exercise: Wrappers for Formatting Currency 77

	4.2.6	Idea 3: Auto-Guessing Defaults	77
	4.2.7	Exercise: Providing Better Defaults for `write.csv`	78
	4.2.8	Idea 4: Split Functionality into Many Functions	79
	4.2.9	Exercise: Decomposing the `quantile` Function	80
	4.2.10	Cyclomatic Complexity	80
	4.2.11	How to Reduce Cyclomatic Complexity	83
	4.2.12	Exercise: Calculating Leap Years	83
4.3	Summary		84

5 Integrating Testing into Packages — **85**

5.1	How to Make an R Package		85
	5.1.1	Prerequisites	85
	5.1.2	The Package Directory Structure	86
	5.1.3	Including Tests in Your Package	86
5.2	Case Study: The *hypotenuser* Package		87
5.3	Checking Packages		90
	5.3.1	Exercise: Make a Package with Tests	91
5.4	Using Version Control, Online Package Hosting, and Continuous Integration		92
	5.4.1	Version Control with *git*	92
	5.4.2	Online Project Hosting	92
	5.4.3	Continuous Integration Services	93
5.5	Testing Packages on CRAN		95
	5.5.1	Testing Packages with r-hub	96
5.6	Calculating Test Coverage Using *coveralls.io*		97
5.7	Summary		98

6 Advanced Development-Time Testing — **99**

6.1	Code with Side Effects		99
	6.1.1	Exercise: Writing a Test That Handles Side Effects	101
6.2	Testing Complex Objects		101
	6.2.1	Exercise: Testing a Complex Object	104
6.3	Testing Database Connections		104
	6.3.1	Option 1: Mock the Connection	106
	6.3.2	Option 2: Mock the Connection Wrapper	107
	6.3.3	Option 3: Mock the Specific Query Functions	108
	6.3.4	Summarizing the Pros and Cons of Each Database Method	110
6.4	Testing Rcpp Code		110
	6.4.1	Getting Set Up to Use C++	111
	6.4.2	Case Study: Extending the *hypotenuser* Package to Include C++ Code	114
	6.4.3	Exercise: Testing an *Rcpp* Function	117
6.5	Testing Write Functions		118
	6.5.1	Exercise: Writing INI Configuration Files	120

6.6   Testing Graphics   120
      6.6.1   Generating the Report from the *markdown*   124
      6.6.2   Including Graphics Tests in Packages   124
      6.6.3   Exercise: Write a Graphics Test Report   124
6.7   Summary   125

7   Writing Your Own Assertions and Expectations   **127**
7.1   The Quick, Nearly Good Enough Option   127
7.2   Writing Scalar Predicates   128
      7.2.1   Exercise: Writing a Custom Scalar Predicate   129
      7.2.2   Writing Type Predicates   129
7.3   Writing Scalar Assertions   130
      7.3.1   Exercise: Writing a Custom Scalar Assertion   131
7.4   Writing Vector Predicates   131
      7.4.1   Exercise: Writing a Vector Predicate   132
7.5   Writing Vector Assertions   133
      7.5.1   Exercise: Writing a Vector Assertion   134
7.6   Creating Custom Expectations   135
      7.6.1   Exercise: Create a Custom Expectation   137
7.7   Summary   137

A   Answers to Exercises   **139**
A.1   Preface   139
      A.1.1   Exercise: Are You Ready?   139
A.2   Chapter 2   139
      A.2.1   Exercise: Using Predicates and Assertions   139
      A.2.2   Exercise: Examining an Object   141
      A.2.3   Exercise: Inspecting Files and Directories   142
      A.2.4   Exercise: Testing Properties of Matrices   143
      A.2.5   Exercise: Explore Your R Setup   143
      A.2.6   Exercise: Checking Customer Data   143
      A.2.7   Exercise: Calculating the Harmonic Mean   147
A.3   Chapter 3   149
      A.3.1   Exercise: Using `expect_equal`   149
      A.3.2   Exercise: Using `expect_error`   150
      A.3.3   Exercise: Using `expect_output`   151
      A.3.4   Exercise: Find More Tests for `square_root`   151
      A.3.5   Exercise: Testing the Return Type of Replicates . . .   152
A.4   Chapter 4   153
      A.4.1   Exercise: Reducing duplication   153
      A.4.2   Exercise: Outsourcing Argument Checking   155
      A.4.3   Exercise: Wrappers for Formatting Currency   156
      A.4.4   Exercise: Providing Better Defaults for `write.csv` . .   157
      A.4.5   Exercise: Decomposing the `quantile` Function   157
      A.4.6   Exercise: Calculating Leap Years   158

A.5   Chapter 5   159
    A.5.1   Exercise: Make a Package with Tests   159
A.6   Chapter 6   162
    A.6.1   Exercise: Writing a Test That Handles Side Effects . .   162
    A.6.2   Exercise: Testing a Complex Object   162
    A.6.3   Exercise: Testing an *Rcpp* Function   163
    A.6.4   Exercise: Writing INI Configuration Files   165
    A.6.5   Exercise: Write a Graphics Test Report   167
A.7   Chapter 7   167
    A.7.1   Exercise: Writing a Custom Scalar Predicate   167
    A.7.2   Exercise: Writing a Custom Scalar Assertion   168
    A.7.3   Exercise: Writing a Vector Predicate   169
    A.7.4   Exercise: Writing a Vector Assertion   169
    A.7.5   Exercise: Create a Custom Expectation   170

Bibliography   **173**

Concept Index   **177**

Package Index   **179**

Dataset Index   **181**

People Index   **183**

Function Index   **185**

Preface

The problem with programming is that you are always one typo away from having written something silly. Likewise with data analysis, a small mistake in your model can give you a big mistake in your results. Combining the two disciplines means that it is all too easy for a missed minus sign or a confusion about a unit[1] to generate a false prediction that you don't spot until after your company has spent a load of money, or the journal accepted your paper.

R is designed to help you do data analysis fast. The high-level commands and features like dynamic typing mean that mostly you can just focus on solving statistical problems, and everything magically just works. Unfortunately, there are a few quirks in the language that occasionally result in obscure bugs and lots of swearing.

This book is about how to write R code that results in fewer bugs (and maybe less swearing). Testing is the only way to be sure that your code, and your results, are correct.

About This Book

Working with R can be a bit daunting. You need statistics skills to solve your data problems, but you also need programming skills to make the language work for you. In fact, a whole new job title – data scientist – was invented to describe people who have to do both jobs. wthou This book helps you hone your skills on the programming side of things. So the first bit of good news is that there are no hard statistical bits in the book. The second bit of good news is that the programming isn't too hard either.

Testing code is conceptually easy; the hard part is developing habits to incorporate testing into your workflow. That means that this book is filled with lots of exercises to help you practise. **Attempt all the exercises!** It really is the best way to become a proficient tester.

Some of the exercises assume basic statistical knowledge – you are assumed to understand the concepts of mean and standard deviation, for example. You

---

[1] Space enthusiasts may recall the Mars Climate Orbiter, which disintegrated in the Martian atmosphere after some software calculated its thrust in US units rather than the metric units it was expecting.

are also assumed to have some basic R skills. For example, you should know how to write a function, know what common variable types like `numeric` vectors and `data.frames` are, and be able to do basic manipulations with them. If you haven't used R before, then put this book to one side for a while, and read *Learning R* [4] (also by me) first.

Exercise: Are You Ready?

To make sure your R skills are good enough that you will find reading this book comfortable, write a function that accepts a data frame and returns the number of numeric columns in the input.

What's in This Book

The first four chapters should be read by everyone. After reading them, you should feel confident enough to start including testing in your own work.

- Chapter 1, Introduction, sets out the concepts of run-time testing and debug-time testing, with some motivating examples.

- Chapter 2, Run-Time Testing with assertive, gives a tour of my *assertive* [8] package for run-time testing.

- Chapter 3, Development-Time Testing with testthat, gives a tour of Hadley Wickham's *testthat* [23] package for development-time testing.

- Chapter 4, Writing Easily Maintainable and Testable Code, applies computer science knowledge of how to make your code more easily maintainable and testable to R programming.

  The next three chapters are primarily for people developing R packages.

- Chapter 5, Integrating Testing into Packages, explains how to include unit tests in your packages. This is suitable for all package developers.

- Chapter 6, Advanced Development-Time Testing, looks at some specialized unit testing scenarios such as testing databases, *Rcpp* [10] code and graphics. Pick the scenarios that are relevant to your work.

- Chapter 7, Writing Your Own Assertions and Expectations, explains how to create your own assertions for *assertive*, and your own expectations for *testthat*. This is a niche chapter for people interested in extending *assertive* or *testthat*.

Things You Need before You Begin

The obvious thing is that you need a copy of R; I'm leaving this as an exercise for you; if you can't figure out how to install R, then it's probably best that you back up a couple of steps and read *Learning R* first.

There are lots of great integrated development environments (IDEs) for R, and I recommend that you use one. The four main choices are:

RStudio This is arguably the best IDE for working only with R. It's made by the same people who created a lot of the packages mentioned in this book, including *devtools*, *testthat*, and *knitr*, and it really benefits from integration with these packages.

Architect This is a remix of Eclipse, with plugins for R integration. You get excellent cross-language capabilities (in case you fancy a bit of Python programming too), and the editor is a little more sophisticated than the one in RStudio. The learning curve is somewhat steeper than that of RStudio.

emacs + ESS You probably already know if you like emacs. It's got a learning curve as steep as a cliff face but power-users swear by it. Also the key chords build up muscles in your little fingers, which is useful if you are a rock climber.

R Tools for Visual Studio Microsoft has recently discovered R in a big way, and has a plugin for using R with the venerable Visual Studio IDE. Visual Studio in mature and slick, but the R tools are new and not as fully featured as the competition[2].

In order to follow along with all the examples and exercises, you also need to install the following packages. In alphabetical order:

- *assertive*

- *BiocCheck*

- *caret*

- *covr*

- *cyclocomp*

- *data.table*

- *devtools*

- *digest*

- *dplyr*

- *ggplot2*

---

[2]Unless you've just bought this book in a thrift store and it's 2023. Hi from 2016!

- *installr*

- *knitr*

- *magrittr*

- *plyr*

- *Rcpp*

- *rebus.datetimes*

- *roxygen2*

- *RPostgreSQL*

- *runittotestthat*

- *sig*

- *testthat*

- *withr*

Additional Resources

The exercises are available online as Jupyter notebooks. You can complete the exercises interactively in a browser, avoiding the need for installing any software. You need to register for a free Anaconda cloud account to access these workbooks.

Exercises only:

https://anaconda.org/richierocks/testing-r-code-exercises-questions-only/notebook

Exercises and answers:

https://anaconda.org/richierocks/testing-r-code-exercises-questions-and-answers/notebook

The package that is created in Chapters 5 and 6 is available from https://bitbucket.org/richierocks/hypotenuser

Acknowledgments

Thanks to my editors Rob Calver and Rebecca Davies for their support in creating this book. Colin Gillespie and two mystery reviewers (you know who you are) gave some great feedback that made this book much stronger. Marcus Fontaine produced the book and provided some timely technical help with

LaTeX. Karen Simon found a near infinite number of typos and grammatical errors. Thanks are also due to the developers of the R packages used here. This book would be half empty without the work of Hadley Wickham's *testthat*, and it would have been much harder to write without Yihui Xie's *knitr* package.

Lastly, but certainly not leastly, thanks go to Janette Cotton, for not complaining about me avoiding household chores while I've been writing.

1

Introduction

OK, here we go! This chapter explains about the two ways of working with R (*interactively* and *programmatically*), and the two types of testing (*run-time* and *development-time*) that you'll be reading about in the rest of the book, and why you need them.

1.1 Interacting vs. Programming

R provides an excellent environment for interacting with your data. The high-level commands mean that with one or two lines of code you can run a model or draw a plot, and find out something new about your dataset. Since you get instant feedback, it's usually fairly easy to spot problems with your code. For this sort of R usage, formal testing techniques aren't so important.

Unfortunately, interactivity doesn't scale well. If you have ten thousand datasets to analyze, you simply don't have time to interact with each one. In this case, you have to rely more on your programming skills. Ideally, you can just loop over each dataset, and they will all behave in the same way. That's the best case scenario. What you'll probably find in practice is that a few of the datasets have some weird edge cases in them. Maybe some missing values, or negative numbers that shouldn't be, or times that are in the future. Since each dataset gets less attention from you, it's very easy for subtle problems to creep into your analysis and ruin your results. If you're not careful, you won't even know that your results are wrong. For these data analyzes at scale, it is vital to put more effort into testing your code.

Another case where testing is incredibly important is when you are providing code for other people to use. For example, if you are developing an R package that will be made publicly available (perhaps on CRAN or Bioconductor), then you need to be sure that your code works properly. If you waste other people's time with scrappy code, your reputation will be shot. The tricky part is that users will almost certainly use your code in ways that you didn't imagine, and on datasets in different formats. On top of this, they'll be running their code under different operating systems. While R is mostly consistent across OSes, there are some areas such as file paths and locales where the behavior is necessarily different.

> If you are doing large scale data analysis, or writing code for other people, then you need to do more testing.

1.2   Two Kinds of Testing

This book deals with two different kinds of testing.

development-time testing is perhaps what many people think of when they think about testing.[1] In the context of R, it usually means writing *unit tests*. These are tests that make sure that your functions give the answer that you expect, or throw the error that you expect. It's called development-time testing because that's when it happens: you write some code, then you write some tests to make sure it works correctly.[2]

The R package we'll use for development-time testing is called *testthat*.

run-time testing happens when a user (possibly you) runs your code. When you develop code, you often have a particular use case in mind. This can lead to you assuming things that may not be true when other people actually come to use the software. Or even if they do use the software as you intended, perhaps their data is in a different form, or has some different properties that you didn't anticipate. Run-time tests are known as *assertions*, though since this is a bit jargony, we'll refer to them as *checks* (with *tests* referring to unit tests).

The R package we'll use for run-time testing is called *assertive*, and we'll look at this type of testing first.

> The point of development-time testing is to make sure that you haven't done something stupid. By contrast, the point of run-time testing is to make sure that the user hasn't done something stupid.

1.3   How Run-Time Testing Will Help You

One of R's greatest strengths in an interactive context can also be a bit of a weakness in a programming context. Dynamic typing means that you don't have to explicitly specify the type of all your variables. This is a great time

---

[1] If they ever think about testing, of course.

[2] Some software development systems such as Extreme Programming suggest that you should write your tests before you write your main code. I personally struggle to get my head around this, but test-driven development has many enthusiastic fans. In any case, you shouldn't put off testing too long.

saver, and most of the time, everything magically just works. However, in the event of bad user input, you don't get automatic compiler checks that tell you something has gone wrong.

Consider the `log` function. There are all sorts of bad forms of input that it could take. Suppose you pass a negative number to the function. Since the logarithm of a negative real number doesn't make much sense, you need to decide how to handle the situation. In fact, you have a choice of possible behaviors. The most severe option is to throw an error, explaining that negative numbers are not allowed. This is probably too strict, since there are cases where the log of a negative number is legitimate code. At the other extreme, you could simply return the complex number solution, without informing the user that there was anything amiss. This is usually too lenient; in many fields of data analysis, complex numbers aren't used at all.[3] For example, business people would be quite annoyed if you predicted that their company profits were going to be imaginary! In between the extremes, you could provide a message or a warning explaining the problem.

R's built-in `log` function has two behaviors, depending upon the input type. For a negative `numeric` input, the function returns NaN, generates a mostly useless warning. For a `complex` input, the return value is also complex.

```
log(-1)
```

```
# Warning in log(-1): NaNs produced
```

```
# [1] NaN
```

```
log(-1 + 0i)
```

```
# [1] 0+3.141593i
```

There is no official way of deciding how these bad user inputs should be handled. It depends upon the scale of the software you are creating – in general, larger projects need to be stricter about things running "correctly," so you should err on the side of throwing errors. It also depends a bit on personality. My boss is a big fan of throwing errors, whereas I tend towards trying to fix inputs.

Another form of bad input to `log` is character data. How should `log("10")` be handled, for example?

Here, again, you can throw an error, or you could try to convert the input to be a number. Of course, trying to convert something like `log("a")` is hopeless, but – if you think it would be useful to the user – the previous example could be made to return the number 10.

In R's case, the built-in function will throw an error for character input.

---

[3]On the other hand, complex numbers crop up surprisingly often in physics.

```
log("10")

# Error in log("10"): non-numeric argument to mathematical function
```

I think that this is the right decision here. Although you lose out on the possibility to help the user by automatically fixing some cases of bad input, you also risk hindering the user by not letting them spot the problems in their data. If they are passing character data to a logarithm function, they almost certainly have a problem earlier on in their code, and throwing an error is the best way to alert them to it.

> For bad user inputs, your choice is between fixing the problem and providing a message or warning to explain what you've done, or throwing an error to say what went wrong.

1.4   How Development-Time Testing Will Help You

Suppose we decide to write our own R math package. Since we were looking at the natural logarithm function, let's have a go at writing our own to put in it. Here's a dumb implementation that adds up the first 100 terms in the function's Taylor series.[4]

```
natural_log <- function(x)
{
  n <- 100
  i <- seq_len(n)
  sum(rep_len(c(1, -1), n) * x ^ i / i)
}
```

Now that we have a function, we need to know if it works. You can stare at code for hours and never be quite sure if it's doing the right thing, so to establish if it works, we need to try some examples and test the output of the function against results that we know should be correct. In this case, we can test against R's built-in `log` function. In general, you'll usually need to calculate the result yourself.[5]

Digging out your high-school math, you should recall that natural log of 1 is 0, so that makes a good first test.

---

[4]For the natural log function, this is commonly known as the Newton–Mercator Series.

[5]Or hope that your favourite search engine finds the answer for you. Though calculating things with pen and paper can be surprisingly refreshing.

```
log(1)
```

```
# [1] 0
```

```
natural_log(1)
```

```
# [1] 0.6881722
```

Well, R's built-in function gets the right answer, which is reassuring, but it looks like there's a mistake in ours. Oh yeah, the x term should read x - 1. Let's update it.

```
natural_log <- function(x)
{
  n <- 100
  i <- seq_len(n)
  sum(rep_len(c(1, -1), n) * (x - 1) ^ i / i)
}
```

Now we need to run our test again.

```
natural_log(1)
```

```
# [1] 0
```

Much better – the test passes. But maybe we got lucky, so we should try another number. Let's try 0.5.

```
log(0.5)
```

```
# [1] -0.6931472
```

```
natural_log(0.5)
```

```
# [1] -0.6931472
```

Excellent! It's starting to look promising. We should do at least one more number to be sure that our function is OK though. The log of zero is minus infinity, which should be an interesting test.

```
log(0)
```

```
# [1] -Inf
```

```
natural_log(0)
```

```
# [1] -5.187378
```

Wow! That failed pretty badly – our function wasn't even close. So now we need to update the function and check that it works for an input of zero. But then we also need to re-run all of our previous tests to make sure that we haven't introduced any new bugs.[6] Once we've done that, we need to test bigger numbers, negative numbers, vector inputs, and non-numeric inputs before we can be satisfied that out function is OK.

All this sounds like hard work, and even worse, pretty tedious hard work. Having to do the same tests over and over is so boring that – unless you are a really boring person – you won't be bothered to do it.

Before this puts you off testing for life, note that the whole point of testing software is to take the pain and the repetition out of testing. In fact, the *testthat* package's README file states that "testthat tries to make testing as fun as possible".

> Ideally, development-time tests should be written once and run lots of times.

1.5   Summary

- Run-time testing in R refers to assertions, which help prevent user errors.

- Development-time testing in R refers to unit tests, which help prevent developer errors.

- Both kinds of testing can save you time and mental anguish.

---

[6]Bugs that break things that were previously working are called *regression bugs*. This has nothing to do with linear regression.

2

Run-Time Testing with *assertive*

"Stop! Something is wrong!" In this chapter, you'll learn the best way for your code to tell this to your users, using run-time testing.

2.1 Using Assertions

There are times when it is a good idea to check the state of your variables, to ensure that they have the properties that you think they have. For example, if you have a count variable, you might want to check that it is numeric, that all the values are non-negative, and that all the values are whole numbers.

Base-R has a function called `stopifnot` that lets you perform such checks. The function accepts an arbitrary number of expressions – three in this case. If any of these expressions doesn't evaluate to a logical vector that contains all `TRUE`s, then an error is thrown. (Otherwise there is no side effect of calling this function.)

```
counts <- c(1, 2.2, 3, 4.5)

stopifnot(
  # check the variable is numeric
  is.numeric(counts),
  # check all values are non-negative
  all(counts >= 0),
  # check all values are whole numbers
  isTRUE(all.equal(counts, round(counts)))
)

# Error: isTRUE(all.equal(counts, round(counts))) is not TRUE
```

This is OK, but the code isn't that easy to read (especially that last line). Worse, the error messages that it produces in the event of failure aren't very user-friendly.

Here's the same example again, written in an *assertive* style. The assertions are designed to make your code easier to read, and to return helpful error messages to users in the event of a check failing.

```
suppressPackageStartupMessages(library(assertive, quietly = TRUE))
suppressPackageStartupMessages(library(magrittr, quietly = TRUE))


library(assertive)
assert_is_numeric(counts)
assert_all_are_non_negative(counts)
assert_all_are_whole_numbers(counts)

# Error in eval(expr, envir, enclos): is_whole_number :
#   counts are not all whole numbers (tol = 2.22045e-14).
# There were 2 failures:
#   Position           Value        Cause
# 1            2 2.2000000000000002 fractional
# 2            4                4.5 fractional
```

Hopefully this code is clear enough that when you come back to it in six months time, it is still obvious what you wanted to do. The error message contains a human-readable sentence, followed by information on the values that caused problems, along with their positions and reasons for failure[1].

2.1.1 Chaining Assertions Together with Pipes

Stefan Bache's *magrittr* [1] package contains several *pipe* operators that pass the results of one calculation into another calculation. The forward pipe operator, %>%, passes the results of the left-hand side into the first argument of a function of the right-hand side. That is, you can replace f(x) with x %>% f. This can improve code readability, particularly when you have lots of nested function calls. Piping has been popularised by the dplyr [29] package, whose documentation has many examples of its use.

Here is the previous example, re-written in a piped style. In this case, the name of the variable in the error message has been changed to ., which refers to the object passed from the left-hand side of the pipe.

```
counts %>%
  assert_is_numeric %>%
  assert_all_are_non_negative %>%
  assert_all_are_whole_numbers

# Error in function_list[[1L]](value): is_whole_number :
#   . are not all whole numbers (tol = 2.22045e-14).
```

---

[1]The error ought to appear to originate in the assertion that caused the problem, but the book was written with *knitr*, which messes with the call stack, hence the error appears to come from a different location.

```
# There were 2 failures:
#  Position               Value        Cause
#  1          2 2.2000000000000002 fractional
#  2          4                4.5 fractional
```

2.2   Using Predicates: *is* and *has* Functions

Each of the `assert` functions has an underlying *is* or *has*[2] function. For example, `assert_is_numeric` calls `is_numeric`,[3] `assert_all_are_non_negative` calls `is_non_negative`, and so on.

Some *is* and *has* functions, such as `is_numeric`, return a single logical value.

```
is_numeric(1:6)
```

```
# [1] TRUE
```

```
is_numeric(letters)
```

```
# [1] FALSE
# Cause of failure:  letters is not of class 'numeric';
#   it has class 'character'.
```

When the check passed, `is_numeric` returned `TRUE`, and when it failed, `is_numeric` returned `FALSE` with a message explaining the problem.

R objects can store extra data using *attributes*. For example, all R objects have a length attribute, and many of them have a *names* attribute or *dim* attribute, the latter storing information about the object's dimensions. You can add any attributes you like to an object using the `attr` function.

When the predicate returned `FALSE`, the message was stored in an attribute named `cause`. You can see this more clearly by using the `attributes` function, which lists the variable's attributes.

```
attributes(is_numeric(letters))
```

```
# $cause
# [1] letters is not of class 'numeric'; it has class 'character'.
#
# $class
# [1] "scalar_with_cause" "logical"
```

---

[2]The prefix "has" usually refers to properties of a variable; "is" is used otherwise.

[3]`is_numeric` is just a wrapper to the base-R function `is.numeric`, with some extra functionality for giving a nicer error message when things go wrong.

Where *is* functions return a single value, they have a single corresponding *assert* function prefixed by `assert_`. For example, `is_numeric` is paired with `assert_is_numeric`.

Some *is* and *has* functions, such as `is_non_negative`, work elementwise (rather than returning a single `TRUE` or `FALSE` value for the whole input).

```
is_non_negative(c(10, 1, 0, -1, -10, NA))
```

```
# There were 3 failures:
#   Position Value    Cause
# 1        4    -1 too low
# 2        5   -10 too low
# 3        6  <NA> missing
```

The result shows the position value and cause of failure for each element where the test did not pass. The result is actually a logical vector with a `cause` attribute – a `print` method has been defined to provide the custom output that you just saw. Again this is easier to see if we use `attributes`.

```
attributes(is_non_negative(c(10, 1, 0, -1, -10, NA)))
```

```
# $names
# [1] "10"  "1"   "0"   "-1"  "-10" NA
#
# $cause
# [1]                         too low too low missing
#
# $class
# [1] "vector_with_cause" "logical"
```

This time the `cause` attribute is also vectorised, returning an empty string for the passes and a brief explanation of the problem for the failures.

Where *is* functions return a vector, there are two corresponding *assert* functions, prefixed `assert_all_are`, and `assert_any_are`. For example, `is_non_negative` is paired with `assert_all_are_non_negative` and `assert_any_are_non_negative`.

2.2.1   Exercise: Using Predicates and Assertions

Consider the variable, x:

```
(x <- c(0, 1, Inf, -Inf, NaN, NA))
```

```
# [1]    0    1  Inf -Inf  NaN   NA
```

Write checks that

1. x is a numeric vector.

2. all the elements of x are finite.

3. all the elements of x are not missing.

Try both the predicate ("is") and assertion ("assert") functions.

Hints: You can search inside the *assertive* package for functions that match a particular pattern using, for example, `ls("package:assertive", pattern = "finite")`. `apropos("finite")` is easier to type, but will search all packages on your search path. Also, two of the checks should fail!

2.3   The Virtual Package System

The *assertive* package isn't just a single package. It consists of 15 underlying packages that contain predicates and assertions for specific topics. The *assertive* package itself doesn't contain any of its own content; it merely re-exports functions from the other packages. This design helps distinguish between interactive and programmatic use.

If you are using *assertive* interactively, then you can simply type `library(assertive)` and have access to all the functionality.

By contrast, if you are developing your own packages, then you can pick and choose which of the packages you need, avoiding the larger dependency of *assertive* itself. Figure 2.1 shows which packages import other packages. You need to have *assertive.base* in order to use all the other packages.

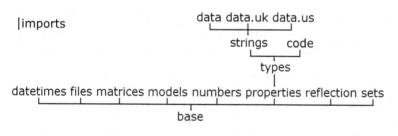

FIGURE 2.1
The hierarchy of *assertive* packages.

2.4   A Tour of the *assertive* Package

This tour covers each of the packages that comprise *assertive*. Some packages may not be relevant to the types of analysis that you do, so feel free to skip some sections. I recommend reading at least the sections on the four packages that are most widely useful for checking input variables: *assertive.base*, *assertive.properties*, *assertive.types*, and *assertive.numbers*.

2.4.1   assertive.base

assertive.base contains the base functionality that you need to do run-time testing. The package contains a few general-purpose predicates: these can be used anywhere, but the error message won't be as clear for users as if you had used a more specific predicate. The package also contains some utility functions that will be discussed in later examples.

There are six predicates that return vectors. is_true returns TRUE whenever the input is TRUE, and FALSE otherwise. That is, it works like x & !is.na(x).

```
x <- c(TRUE, FALSE, NA)
is_true(x)

# There were 2 failures:
#    Position Value    Cause
# 1         2 FALSE    false
# 2         3  <NA>  missing
```

> What do you think the two corresponding assert functions are called?[4]

Similarly, is_false and is_na check for those logical values, and is_not_true, is_not_false, and is_not_na check for the opposite cases.

> Try calling those five other predicates (is_false, etc.) to see their effects on x. Call the corresponding assertions too! Which ones do you think will fail?[5]

Four functions that return a single value. is_identical_to_true works like the base-R function isTRUE; that is, it checks for identical(x, TRUE). The base-R function is a little annoying in that often you don't really care about whether or not the variable you are checking has attributes[6] like names – you just want to know if the result is TRUE or not. is_identical_to_true solves this by giving you the option of stripping any attributes before the check is made.

---

[4]If you said assert_all_are_true, and assert_any_are_true, well done!

[5]In this case, the assert_any assertions will pass and the assert_all assertions will fail.

[6]There is a refresher on attributes in the next section on *assertive.properties*.

```
(x <- c(a = TRUE))

#    a
# TRUE

isTRUE(x)                    # this is a bit weird, right?

# [1] FALSE

is_identical_to_true(x)

# [1] FALSE
# Cause of failure:  x is not identical to TRUE;
#    its value is structure(TRUE, .Names = "a").

is_identical_to_true(x, allow_attributes = TRUE)

# [1] TRUE
```

is_identical_to_false and is_identical_to_na are the equivalent functions for testing single FALSE and NA values.

> What does is_identical_to_na(matrix()) return?[7]

2.4.2   assertive.properties

assertive.properties contains functionality for checking the properties of variables, such as length, class, or attributes. Attributes are extra bits of data strapped on to a variable; the most common ones you should have come across are names and dim (for the object's dimensions). You can retrieve or add arbitrary attributes using the attr function.

has_names checks a variable to see if it has at least one non-empty name.

```
x <- 1:5
has_names(x)
names(x) <- character(5)
has_names(x)
names(x) <- month.abb[1:5]
has_names(x)


# [1] FALSE
# Cause of failure:  The names of x are NULL.
# [1] FALSE
```

---

[7]You should see that the check passes only when allow_attributes = TRUE.

```
# Cause of failure:   The names of x are all empty.
# [1] TRUE
```

Similarly, has_rownames, has_colnames, and has_dimnames check for those particular kinds of names specific to data.frames, matrices, and arrays.

---

stack.x is a built-in dataset of chemical plant operation data. (See the ?stack.x help page for more information.) Try has_names, has_rownames, has_colnames, and has_dimnames on this dataset. Which checks pass?[8]

---

is_scalar checks variables to see if they have length one. In the following examples, recall that numeric creates a numeric vector of the length of the input, defaulting to zero length.

```
is_scalar(numeric())

# [1] FALSE
# Cause of failure:  numeric() has length 0, not 1.

is_scalar(numeric(1))

# [1] TRUE

is_scalar(numeric(2))

# [1] FALSE
# Cause of failure:  numeric(2) has length 2, not 1.
```

You can also use it to check for variables with one element. For most variable types, this is exactly the same thing. It isn't true of recursive variable types like lists and data.frames. In the next example, the list has length one, since there is only one top-level element, but that element contains five numbers.

```
l <- list(a = 1:5)
is_scalar(l)

# [1] TRUE

is_scalar(l, metric = "elements")

# [1] FALSE
# Cause of failure:  l has 5 elements, not 1.
```

---

[8]The dataset has column names (and hence dimension names) but not names or row names.

There are lots more functions for checking lengths and dimensions: is_empty checks for objects with length zero (or zero elements), and is_non_scalar and is_non_empty check for the opposite cases. is_of_length, has_elements, and is_of_dimension provide more general variable size checks. has_dims, has_rows, and has_cols check for the dimensions, rows, and columns, respectively. are_same_length and have_same_dims check for two variables having the same length/dimensions.

Are functions scalar?[9]

The previous examples looked at specific attributes (names and dimensions). You can also check for the presence of arbitrary attributes using has_attributes, which is vectorized over its second argument. has_no_attributes and has_any_attributes check for the presence or absence of any attributes at all.

```
has_attributes(data.frame(), c("names", "class", "dim"))

# There was 1 failure:
#    Position Value    Cause
# 1         3   dim no attr

has_no_attributes(data.frame())

# [1] FALSE
# Cause of failure:  data.frame() has the attributes
#    names, row.names, class.

has_any_attributes(1:5)

# [1] FALSE
# Cause of failure:  1:5 has no attributes.
```

There are three attributes that all data frames should have. What are they? Check that an empty data frame, data.frame(), has these attributes.[10]

---

has_no_duplicates checks that a vector has no duplicates. This is particularly useful for things like ID columns of datasets, which often have to be unique.has_duplicates checks the opposite.

---

[9]All functions have length one, and hence are scalar with the default metric. The number of elements is usually more than one though.

[10]The three compulsory attributes are names, row.names, and class.

```
(x <- c(1, 2, 3, 2, 4, 1))

# [1] 1 2 3 2 4 1

has_no_duplicates(x)

# [1] FALSE
# Cause of failure:   x has duplicates at positions 4, 6.

has_duplicates(1:5)

# [1] FALSE
# Cause of failure:   1:5 has no duplicates.
```

There are a few more property-related predicates in this package, including is_null, and is_not_null for checking whether an object is or isn't NULL; is_atomic and is_recursive for checking recursivity; is_vector that checks for vectors; and is_unsorted that checks for unsorted vectors.

2.4.3   assertive.types

assertive.types contains functionality for checking types of variables. Most of these wrap functions in base-R. You've already seen is_numeric that wraps is.numeric, and there are many more equivalent functions for other variable types, from the common things like is_data.frame through to rarer things like is_qr and is_relistable, and types from popular non-base packages like is_data.table.

Perhaps more interesting are the functions that combine checks for types with is_scalar from *assertive.properties*. These let you check for a single number (via is_a_number), or a single string (is_a_string), or a single logical value (is_a_bool).

```
is_a_number(1.23)

# [1] TRUE

is_a_number("1.23")

# [1] FALSE
# Cause of failure:   "1.23" is not of class 'numeric';
#   it has class 'character'.

is_a_number(1:23)

# [1] FALSE
# Cause of failure:   1:23 has length 23, not 1.
```

> Does the built-in variable **LETTERS**, containing the upper-case Roman alphabet, pass the check for `is_a_string`?[11]

2.4.4 assertive.numbers

assertive.numbers lets you check that your numeric vectors are as they should be. We've already used some of its functionality, like `is_non_negative` and `is_whole_number`.

`is_non_negative` is part of a set of functions for checking that numbers are in a particular range. The most general of these is `is_in_range`, but the more specialized wrappers are typically more useful. There are four predicates for checking whether or not values are greater than or less than zero: `is_negative`, `is_positive`, `is_non_positive`, and the previously mentioned `is_non_negative`.

You can also test for proportions (in the range zero to one). This function allows you to choose whether the end-points are considered to be in range.

```
x <- c(-0.1, 0, 0.5, 1, 1.1)
is_proportion(x)    # 0 <= x <= 1

# There were 2 failures:
#    Position Value     Cause
# 1         1  -0.1   too low
# 2         5   1.1  too high

is_proportion(       # 0 < x < 1
  x,
  lower_is_strict = TRUE,
  upper_is_strict = TRUE
)

# There were 4 failures:
#    Position Value     Cause
# 1         1  -0.1   too low
# 2         2     0   too low
# 3         4     1  too high
# 4         5   1.1  too high
```

`is_percentage` works in the same way for the range zero to one hundred.

> How might you check for valid angles, specified as a number between zero and two pi?[12]

---

[11]LETTERS doesn't contain a string, since it has length 26.

Rather than checking a range, you can also check for comparisons against a single number, with predicates for the relational operators. Checking for equality is done with a default tolerance of about `2.2e-14`, which is just enough to make sure that floating-point arithmetic errors shouldn't make the same number appear different.[13]

```
eps <- 50 * .Machine$double.eps
x <- 123.456 + c(-10 * eps, -eps, 0, eps, 10 * eps)
is_equal_to(x, 123.456)

# There were 2 failures:
#   Position              Value
# 1          1 123.45599999999989
# 2          5 123.45600000000012
#                                          Cause
# 1 not equal to 123.456 (tol = 2.22045e-14);
#    abs. difference = 1.13687e-13
# 2 not equal to 123.456 (tol = 2.22045e-14);
#    abs. difference = 1.13687e-13

is_equal_to(x, 123.456, tol = 0)

# There were 4 failures:
#   Position              Value
# 1          1 123.45599999999989
# 2          2 123.45599999999999
# 3          4 123.45600000000002
# 4          5 123.45600000000012
#                                                                         Cause
# 1 not equal to 123.456 (tol = 0); abs. difference = 1.13687e-13
# 2 not equal to 123.456 (tol = 0); abs. difference = 1.42109e-14
# 3 not equal to 123.456 (tol = 0); abs. difference = 1.42109e-14
# 4 not equal to 123.456 (tol = 0); abs. difference = 1.13687e-13

is_not_equal_to(x, 123.456)

# There were 3 failures:
#   Position              Value                                      Cause
# 1          2 123.45599999999999 equal to 123.456 (tol = 2.22045e-14)
# 2          3            123.456 equal to 123.456 (tol = 2.22045e-14)
# 3          4 123.45600000000002 equal to 123.456 (tol = 2.22045e-14)
```

Similarly, there are wrappers for is_greater_than, is_less_than, is_greater_than_or_equal_to, and is_less_than_or_equal_to.

Many more numeric checks are possible; for example, is_real and

---

[12]Try is_in_range(x, 0, 2 * pi, upper_is_strict = TRUE).

[13]The actual tolerance is 100 * .Machine$double.eps, the definition of which is slightly technical but well explained on the ?.Machine help page.

is_imaginary check complex numbers to see if they lie on the real or imaginary axes, and there are predicates for checking infinities and NaN, as well as odd and even numbers.

2.4.5 Exercise: Examining an Object

Harman23.cor is one of the datasets supplied with base-R. It contains a correlation matrix of physical measurements (height, weight, etc.) of seven to seventeen year old girls. Type the name of the dataset to see it.

Perform the following checks on the structure of Harman23.cor:

- The dataset is a list.

- The dataset has length three.

- The "cov" element of the dataset is a numeric matrix with both row names and column names.

- All values of the "cov" element are between zero and one.

- All values of the "center" element are zero.

2.4.6 assertive.files

assertive.files contains functionality for working with files and connections.

is_existing_file wraps base-R's file.exists to check if character vectors point to files on your machine or network. It solves an annoying quirk of the base-R function under Windows where trailing slashes can confuse the function into not finding files that exist.

```
r_executable <- file.path(R.home("bin"), "R") # "R.exe" under Windows
is_existing_file(r_executable)

# /Library/Frameworks/R.framework/Resources/bin/R
#                                                          TRUE
```

is_dir checks for directories, and you can check for the permissions of a file or directory using is_readable_file, is_writable_file, and is_executable_file. In the following example, we look at the contents of the base package.

assertive.files also contains functions for checking connections. This is fairly advanced stuff that you shouldn't usually need. Perhaps the most common use case is with functions that read or write data from files. Many of these functions accept either a string giving a path to a file, or a file connection. So a useful piece of logic is

```
assert_any_are_true(
  c(
    is_a_string(x),
    is_file_connection(x) && is_readable_connection(x)
  )
)
```

2.4.7 Exercise: Inspecting Files and Directories

You can retrieve the contents of the root of the base package using:

```
contents_of_base_pkg_dir <- dir(
  system.file(package = "base"),
  full.names = TRUE
)
```

Check this character vector to see which files exist, are directories, and are readable/writable/executable.

2.4.8 assertive.strings

assertive.strings contains functionality for working with, as you might expect, strings.

There's a function in base-R, nzchar ("non-zero characters"), for checking for non-empty strings. is_non_empty_character wraps this. Since nzchar is a bit weird when it comes to missing strings – NA is considered to be two characters long – you also get is_non_missing_nor_empty_character to exclude missing strings. These assertions are particularly useful for checking columns of data that must contain text data everywhere.

```
x <- c("a", "", NA)
is_non_empty_character(x)

# There was 1 failure:
#   Position Value Cause
# 1        2       empty

is_non_missing_nor_empty_character(x)

# There were 2 failures:
#   Position Value   Cause
# 1        2         empty
# 2        3  <NA> missing
```

is_empty_character and is_missing_or_empty_character complete the set of situations about whether missing and empty strings are allowed. There are also functions that combine the above functions with is_scalar, checking for a single empty, missing (or not) string.

Which elements would fail in the previous example if you called is_empty_character and is_missing_or_empty_character?[14]

Other assertions in the package include is_numeric_string, which checks for strings that contain numbers in a form that R understands, and is_single_character that checks for strings that contain exactly one character.

2.4.9 assertive.matrices

Can you guess what data type *assertive.matrices* provides checks for? Yup, it's matrices.

There are several methods for identifying special matrix types. For example you can check for identity matrices, diagonal matrices, and zero matrices. The functions for these checks allow a tolerance argument that works in the same way as those discussed in the *assertive.numbers* package section.

```
(m <- matrix(c(1, eps, 0, 1), nrow = 2))

#                [,1] [,2]
# [1,] 1.000000e+00    0
# [2,] 1.110223e-14    1

is_identity_matrix(m)

# [1] TRUE

is_identity_matrix(m, tol = 0)

# [1] FALSE
# Cause of failure:  m contains non-zero elements:
#    row col        value
# 1    2   1 1.110223e-14
```

Notice in the above results, when you have a failure, the locations and values of the offending elements are shown.

Other matrix properties can be checked via is_diagonal_matrix, is_symmetric_matrix, is_square_matrix, is_upper_triangular_matrix, is_lower_triangular_matrix, and is_zero_matrix.

---

[14]The first and third elements are not empty and would fail the is_empty_character check. Only the first element would fail the is_missing_or_empty_character check.

2.4.10   Exercise: Testing Properties of Matrices

A correlation matrix should be symmetric and contain real values between minus one and one (including these end-points). How would you check that a matrix passes these criteria?

Test against the correlation matrix generated by `cor(longley)`, which uses the `longley` macroeconomic dataset from the *datasets* package.

2.4.11   assertive.sets

assertive.sets is a small package that allows you to do set comparisons. While R doesn't have a native `set` variable type (you need to use the excellent `sets` package for this), if you use atomic vectors and ignore the order of the elements, then they behave like sets do.

```
x <- c(1, 3, 5, 4, 2)
y <- c(6, 1, 4, 2, 3, 5)
is_subset(x, y)

# [1] TRUE

is_superset(x, y)

# [1] FALSE
# Cause of failure:  The element '6' in y is not in x.

are_set_equal(x, y)

# [1] FALSE
# Cause of failure:  c(1, 3, 5, 4, 2) and c(6, 1, 4, 2, 3, 5)
#    have different numbers of elements (5 versus 6).
```

> Check that the built-in `women` dataset contains the columns `weight`, and `height`, in any order[15].

2.4.12   assertive.models

assertive.models is another small package for working with model objects.

`has_terms` checks objects to see if they have a "terms" attribute or component, which is a good indicator of whether they represent a statistical model or not. The following example uses the `chickwts` dataset, which measures the growth rate of chickens given various diets.

---

[15]Try `are_set_equal(colnames(women), c("weight", "height"))`.

```
model <- lm(weight ~ feed, chickwts)
has_terms(chickwts)   # data frames are not models!

# [1] FALSE
# Cause of failure:  chickwts has no terms component nor attribute.

has_terms(model)

# [1] TRUE
```

is_empty_model and its negation is_non_empty_model check model objects to see if they are the empty model or not. Empty models have no intercept and no factors (in this case factors refers to independent variables in the model, not factors that hold categorical variables.) One of my favourite quirks of R is that it has two built-in datasets related to chicken weights; the next example uses the other one.[16]

```
an_empty_model <- lm(weight ~ 0, ChickWeight)
is_empty_model(an_empty_model)

# [1] TRUE

a_model_with_factors <- lm(uptake ~ conc * Type, CO2)
is_empty_model(a_model_with_factors)

# [1] FALSE
# Cause of failure:  a_model_with_factors has factors.
```

Do you think a model with only an intercept counts as an empty model? Try it![17]

2.4.13   assertive.reflection

assertive.reflection lets you check the state of your system: how you are running R, what OS and IDE you are running, and so on.

is_r_devel checks if you are running a development version of R. If you are running a development version of R, then it presumably means that you are doing some sort of development work (perhaps making sure that a package works with the latest version of R). In this case you may want different logic than with regular R usage.

---

[16]If you are a farmer or agricultural data scientist, feel free to submit your chicken weight data to R-Core and see if you can make it three datasets.

[17]Try is_empty_model(lm(uptake ~ 1, CO2)).

```
if(is_r_devel())
{
  # development-only code
}
```

is_r_release, is_r_alpha, is_r_beta, is_r_release_candidate, is_r_revised, and is_r_patched check for the other build types of R, though they have more niche use cases.

You can also check if you are running R interactively or in batch mode via is_interactive and is_batch_mode. Similarly, is_r_slave checks if you are running a slave instance of R – R runs in this mode parallel computing nodes, and when it is building or checking packages.

There are functions for checking the operating system: is_windows, is_osx, is_linux, is_bsd, is_solaris, and the more general is_unix. For Windows and OS X, you can also check particular versions of the operating system via, for example, is_windows_10 or is_osx_el_capitan.

Some IDEs have slightly different capabilities. For example, RStudio and Architect override the default graphics window functionality. You can check for the IDE using is_rstudio, is_architect, and is_revo_r. For RStudio, you can further check if you are running the desktop or server version via is_rstudio_desktop, and is_rstudio_server.

2.4.14 Exercise: Explore Your R Setup

Run the examples for is_r, is_interactive, and is_windows to see a variety of checks on how you are running R.

Perform a check to see if your version of R is up to date.

2.4.15 assertive.datetimes

assertive.datetimes helps you check dates and times.

is_date_string lets you check if a character vector contains strings in a particular date format (ISO 8601 format by default). The format strings are specified in the same way as the base-R function strptime. See that function's help page for more details. In the following example %d means the day of the month, %b means the month of the year, written as a three-letter abbreviation, and %Y means the year, including the century.

```
is_date_string(c("21JUL1954", "wednesday"), "%d%b%Y")

# There was 1 failure:
#   Position    Value      Cause
# 1         2 wednesday bad format
```

is_before and is_after check whether date-time objects (in the three base-R types: Dates, POSIXlts, POSIXcts, or things coercible to the latter) are before or after some time point. `is_in_future` and `is_in_past` provide the same service, tested against the current time. `is_in_past` is particularly useful for checking customer data, to make sure that account opening dates or dates of birth haven't erroneously been recorded in the future.

```
x <- c("2000-02-29", "2029-04-13")
is_in_past(x)

# Warning: Coercing x to class 'POSIXct'.

# There was 1 failure:
#   Position      Value      Cause
# 1         2 2029-04-13  in future
```

2.4.16   assertive.data, assertive.data.us, and assertive.data.uk

There are three packages for checking more complex data types. *assertive.data* contains checks for data types that aren't country specific, such as credit card numbers and email addresses. *assertive.data.us* and *assertive.data.uk* provide checks for data types specific to the United Stated and United Kingdom, respectively, such as phone numbers and zip/post codes.

The credit card checker, `is_credit_card_number`, checks for specific number patterns that are valid, as well as ensuring that the card's checksum is correct. The following example shows Visa credit card numbers, which are 16 digits beginning with a 4.

```
x <- c("4012888888881881", "1012888888881881", "4012888888881882")
is_credit_card_number(x, type = "visa")

# There were 2 failures:
#   Position            Value           Cause
# 1         2 1012888888881881      bad format
# 2         3 4012888888881882 bad checkdigit
```

Other data types supported include US social security numbers, UK national insurance numbers, UK car license plates, honorifics ("Mr" and "Mrs"), IP addresses, ISBN book numbers, Chemical Abstract Service (CAS) numbers for molecules, hex colors.

If you live in the US or UK, check your own zip/post code. (If you live elsewhere, pretend that you've moved in with Queen Elizabeth II at Buckingham Palace, post code `SW1A 1AA`.)

2.4.17   Exercise: Checking Customer Data

The `customer_data` dataset contains (made-up) personal data from some customers. Import the dataset.

> The `customer_data` dataset is avilable from bitbucket in the
> repository for the *hypotenuser* package featured in later chapters.
> Download it from
>
>   https://bitbucket.org/richierocks/hypotenuser/downloads/customer_data.csv.

```
customer_data <- read.csv(
  "customer_data.csv",
  stringsAsFactors = FALSE
)
str(customer_data)
```

```
# 'data.frame': 20 obs. of  7 variables:
# $ Id         : int  1 2 3 4 5 ...
# $ Title      : chr  "MR" "MRS" ...
# $ FirstName  : chr  "Vaughn" "Shi" ...
# $ LastName   : chr  "F" "Ma" ...
# $ DateOfBirth: chr  "1927-01-09" "1957-10-12" ...
# $ Telephone  : chr  "(01294) 907358 " "(0114) 229 71370" ...
# $ Postcode   : chr  "KA12 8SE " "S7 1FF " ...
```

Use the assertive package to find bad data points that may require cleaning.
In particular, check

- The "ID" field should contain positive whole numbers with no duplicates.

- The "Title" field should be a character vector of honorifics.

- The "FirstName" field should be a character vector of strings that aren't
  missing or empty.

- The "LastName" field should be a character vector of strings of a sensible
  length.

- The "DateOfBirth" field contains dates in the year-month-day format,
  %Y-%m-%d, and that they imply ages between 18 and 100.

- The "Telephone" field contains valid UK telephone numbers.

- The "Postcode" field contains valid UK postcodes.

2.4.18 assertive.code

assertive.code contains some checks on R code.

   is_valid_variable_name checks if a character vector contains valid variable names.

```
is_valid_variable_name(c("x_y.z", "x y z"))

# There was 1 failure:
#   Position Value      Cause
# 1        2 x y z bad format
```

> How long can a variable name be before R considers it invalid?[18]

   is_binding_locked checks a variable to see if its binding has been locked, that is, it has been made read-only. You can use locked variables for constants. (The **pryr** package has some more tools for working with constants.) In the following example, R locks variable in a particular environment, so for convenience we create an environment to store the constant in.

```
e <- new.env()
e$TWO_PI <- 2 * pi
is_binding_locked(TWO_PI, e)

# [1] FALSE
# Cause of failure:  TWO_PI is not locked (read-only)
#   in environment<0x103091f48>.

lockBinding("TWO_PI", e)
is_binding_locked(TWO_PI, e)

# [1] TRUE
```

   is_if_condition checks if a variable is suitable for use as a condition in an if block. That is, it is exactly TRUE or FALSE. It is very similar to is_a_bool, but missing values are not allowed.

```
is_if_condition(c(TRUE, FALSE))

# [1] FALSE
# Cause of failure:  c(TRUE, FALSE) has length 2, not 1.
```

---

[18]Variable names up to 10000 characters are OK. (9937 more than you are allowed in MATLAB!) Try assert_all_are_valid_variable_names(paste(rep("a", 10001), collapse = "")).

2.5   Controlling Severity

The assertions all take an argument named `severity`, which controls how severe the consequences of failure are. By default, the value is `"stop"`, indicating an error should be thrown. The other possibilities are `"warning"`, `"message"`, or `"none"`. This example results in a warning rather than the usual error.

```
assert_all_are_nan(-1:1 / 0, severity = "warning")

# Warning: is_nan : -1:1/0 are not all NaN.
# There were 2 failures:
#  Position Value    Cause
# 1         1  -Inf a number
# 2         3   Inf a number
```

A failing assertion typically takes a few milliseconds to generate an error, warning, or message. Usually this isn't a problem, but in simulations (where a function may be called billions of times) or performance-critical production environments, every millisecond can be important. In this case, it's tempting to strip your code down to the bare minimum calculation code. This will improve the speed at which your code runs, but makes debugging difficult. A better option is to keep the assertions in for debugging, but set the severity to `"none"` when you need speed. This option performs almost no work, so it runs very quickly.

> Set `severity = "none"` for production environments where code performance is critical.

2.6   Fail Early, Fail Often

You've seen quite a few examples of different assertions now, but you may be wondering where to use them in your own code. It turns out that there's a very well established rule for what to do. As Jim Gray said (back in 1985!):

Make each module fail fast – either it does the right thing or it stops.

"Fail fast" means that as soon as you can detect that there is a problem, you should correct that problem or throw an error. In practical terms, that mostly means that assertions belong close to the start of your functions.

More recently, this idea has been updated in the phrase "fail early, fail often." The "fail often" part means that you need lots of checks to ensure the integrity of your code. This is especially true of R code: the flexibility of the language gives you a lot of scope for things going wrong.

2.7   Case Study: Calculating the Geometric Mean

Perhaps the most important use of assertions is to check the inputs to functions. In this example, we'll write a function to calculate geometric means, and then update it with assertions to see how it improves.

R doesn't have a built-in function for calculating geometric means,[19] so let's define one here. Recall that the geometric mean is the exponent of the arithmetic mean of the logarithm of the data.

```
geomean <- function(x, na.rm = FALSE)
{
  exp(mean(log(x), na.rm = na.rm))
}
```

In a statically typed language, we could enforce x being a numeric vector. R?s dynamic typing (while mostly helping us be more productive) gives us some rope to hang ourselves with: x and na.rm can be absolutely anything. We need to handle the cases when x is not numeric, or when x contains negative values, or when na.rm is not a single logical value.

The built-in functions exp, mean, and log have some of their own logic for handling bad inputs, and it is possible to simply rely on that logic rather than writing your own. Let's see what happens when we pass a non-numeric value of x to the function.

```
geomean("a")

# Error in log(x): non-numeric argument to mathematical function
```

The error message is OK, but since it is appearing to come from log(x), it isn't totally clear to the user where the problem originates. The *assertive* fix is to include assert_is_numeric(x) in the function. Where should this line go? In accordance with the first clause of the programming principle "fail early, fail often," the assertion belongs at the start of the function.

---

[19]My best guess for the reason for this is that every university Intro to R class needs to teach writing functions, and this is a really convenient first example.

```
geomean2 <- function(x, na.rm = FALSE)
{
  assert_is_numeric(x)
  exp(mean(log(x), na.rm = na.rm))
}
geomean2("a")
```

```
# Error in geomean2("a"): is_numeric : x is not of class 'numeric';
#   it has class 'character'.
```

This makes it clearer where the problem originates; it also gives the user additional feedback about the type of variable that they passed in to the function.

The geometric mean doesn't make any mathematical sense for (real) negative numbers and will return NaN if the input contains any.

```
geomean2(rnorm(20))
```

```
# Warning in log(x): NaNs produced
```

```
# [1] NaN
```

The warning here is not so informative (why were the NaNs produced?), and again it appears to come from `log(x)`. We could be strict and throw an error if there are negative values by adding a call to `assert_all_are_non_negative`. However, the base-R behaviour of returning NaN seems sensible; to replicate this we use the predicate form (`is_non_negative`) and define custom actions based upon its result.

```
geomean3 <- function(x, na.rm = FALSE)
{
  assert_is_numeric(x)
  # Don't worry about NAs here
  if(any(is_negative(x), na.rm = TRUE))
  {
    warning(
      "x contains negative values, ",
      "so the geometric mean makes no sense."
    )
    return(NaN)
  }
  exp(mean(log(x), na.rm = na.rm))
}
geomean3(rnorm(20))
```

```
# Warning in geomean3(rnorm(20)): x contains negative values,
#   so the geometric mean makes no sense.
```

```
# [1] NaN
```

The mean function coerces `na.rm` to be a logical value, warning if the value's length is more than one. In the next example, we generate a numeric vector with some missing values, then pass a silly value of `na.rm`: a numeric vector of length two.

```
x <- rlnorm(20)
x[sample(20, 5)] <- NA
geomean3(x, na.rm = c(1.5, 0))

# Warning in if (na.rm) x <- x[!is.na(x)]: the condition has
#   length > 1 and only the first element will be used

# [1] 1.62527
```

The warning about the length is OK, but again its source is not totally clear for users since it comes from `if` rather than from `geomean3` itself. The coercion to logical happens silently, which isn't ideal.

Again, we could be strict and throw an error when `na.rm` isn't a scalar logical value using `assert_is_a_bool`. (This is a compound assertion checking both the type and the length of the object.) In this case, to replicate the base-R behaviour, we will use some utility functions provided by *assertive*. `use_first` returns the first element of an object, warning if it has length more than one. `coerce_to` converts an object to a different type (in this case to logical) with a warning, using an appropriate as.* function if it exists, or the more general `as` function if it doesn't.

```
geomean4 <- function(x, na.rm = FALSE)
{
  assert_is_numeric(x)
  if(any(is_negative(x), na.rm = TRUE))
  {
    warning(
      "x contains negative values,",
      "so the geometric mean makes no sense."
    )
    return(NaN)
  }
  na.rm <- coerce_to(use_first(na.rm), "logical")
  exp(mean(log(x), na.rm = na.rm))
}
geomean4(x, na.rm = c(1.5, 0))

# Warning: Only the first value of na.rm (= 1.5) will be used.
# Warning: Coercing use_first(na.rm) to class 'logical'.

# [1] 1.62527
```

2.7.1   Exercise: Calculating the Harmonic Mean

The harmonic mean is defined as the reciprocal of the arithmetic mean of the reciprocal of the data, for non-zero inputs. Here's a function to calculate it.

```
harmmean <- function(x, na.rm = FALSE)
{
  1 / mean(1 / x, na.rm = na.rm)
}
```

Update the function to provide better input checking. In particular, you should check that x is numeric and always non-zero, and that `na.rm` is a single logical value. (Whether you want to throw an error for bad input or try to correct the problem is up to your judgement.)

To make sure that your function works as expected, call it with a variety of inputs. Remember to choose bad inputs as well as good ones.

2.8   Alternatives

As with many tasks, there is more than one way of doing it in R. In fact, there are (at least) five other R packages for assertions. As the creator of the *assertive* packages, I naturally think that *assertive* is the best solution, but it's worth taking a quick look at the alternatives. In decreasing order of should-you-consider-using-it:

ensurer is worth a look because it is lightweight and elegant – you don't get any pre-canned assertions, but it's straightforward to create your own. There's also a nice syntax for created type-safe functions, where assertions for checking data types can be automatically generated within a function.

assertr is focussed on checks for data frames. While in an early stage of development, it may provide a complement to *assertive*.[20]

checkmate is more or less a subset of *assertive*, mostly checks for types and files, but it only provides assertions, not the underlying predicates.

tester is another subset of *assertive*, this time only containing predicates but not assertions.

assertthat was the run-time testing sibling to *testthat*, but is no longer under development.

---

[20]It is also possibly transitioning to be built on top of *assertive*, if Tony gets around to it.

2.9   Summary

In this chapter, you have

- Seen how to use predicates and assertions in the *assertive* package.

- Taken a tour of the available functionality in *assertive*.

- Tried using predicates and assertions. (You did attempt the exercises, right?)

3

Development-Time Testing with testthat

You know when the best time to find bugs is? Before you release your code to your users. This chapter explains how to catch bugs using development-time testing.

3.1 Using Unit Tests

The assertions that we used in the run-time testing section are mainly used for checking that your users haven't broken your code. Development-time testing is more about checking that your code gives the right answer in the first place.

This chapter focusses on writing and running unit tests – tests on individual functions. The second verb there is surprisingly important, since being able to easily run tests protects you from a dangerous idiot: future you! I'm sure that in the present, you're a perfectly intelligent person who writes delightful code. Unfortunately, lurking inside every programmer is a future fool who comes back to old code, having forgotten everything about it, and starts changing and breaking things. Having tests in place that you can re-run prevents future you doing too much damage.

Before we get ahead of ourselves, let's take a look at an example.

3.2 The Structure of a Unit Test

Consider a simple function for calculating hypotenuses of right-angled triangles (on a flat surface).

```
hypotenuse <- function(x, y)
{
  sqrt(x ^ 2 + y ^ 2)
}
```

> For simplicity, there are no assertions on the inputs to
> `hypotenuse`. What assertions would you add?

To test this, we need some examples of triangles where we know the answer.
It's best to start simple, so let's use the famous 3-4-5 triangle.

The basic unit test structure consists of a call to `test_that`, which takes
two arguments. The first argument is a string that describes what the test
is for. This enforces documentation of the test, which is really important for
when you come back to it six months later. The second argument contains the
test code itself. Take a look at the code chunk, and we'll go through it line by
line afterwards.

```r
library(testthat)
test_that(
  "hypotenuse, with inputs x = 3 and y = 4, returns 5",
  {
    expected <- 5
    actual <- hypotenuse(3, 4)
    expect_equal(actual, expected)
  }
)
```

It's easier than you might have expected, I hope. Most tests consist of just
three lines of code: declare what you think the answer should be, calculate it,
then test if the two values are equal. Simple.

3.2.1   Exercise: Using `expect_equal`

Write a test for the hypotenuse function, using a 5-12-13 triangle.

Modify the expected value so that the test fails. How much do you need
to change the expected value by before it fails?

Write a test to see what happens when you pass it very large inputs: use `x`
`= 1e300` and `y = 1e300`. Use Pythagorus's theorem to calculate the expected
value. Does the test pass? If not, why not?

Similarly, write a test for the `hypotenuse` function to see if it works with
very small inputs: use `x = 1e-300` and `y = 1e-300`. Does this test pass?
Should this test pass?

3.2.2   How Equal Is Equal?

`numeric` values are floating-point values, which means that small errors can
crop up in computation.[1] To get around this issue, `expect_equal` is built

---
[1]See FAQ on R 7.31, Why doesn't R think these numbers are equal?

upon the base-R function `all.equal`, which allows a small tolerance for error. The exact default amount of error is calculated in a rather complicated way – see the Details section of the `?all.equal` help page for the full details – it is sometimes a relative difference based upon `actual / expected`, and sometimes an absolute difference of `abs(actual - expected)`.

In the previous exercise, you may have spotted a problem with testing small numbers. Although the hypotenuse function underflows and returns zero, since the expected value is very small, `expect_equal` doesn't treat the difference as a failure. You can fix this by enforcing relative errors.

To enforce absolute differences, pass `scale = 1` to `expect_equal`. To enforce relative differences, pass `scale = expected` to `expect_equal`.

```
test_that(
  "with absolute differences the problem isn't spotted",
  {
    expected <- sqrt(2) * 1e-300
    actual <- hypotenuse(1e-300, 1e-300)
    expect_equal(actual, expected, scale = 1)
  }
)
test_that(
  "with relative differences we pick up on the error",
  {
    expected <- sqrt(2) * 1e-300
    actual <- hypotenuse(1e-300, 1e-300)
    expect_equal(actual, expected, scale = expected)
  }
)
```

> If your equality tests are failing, try passing the `scale` argument to enforce relative or absolute differences – if you know which one is being tested for, it is easier to reason about.

3.3 Testing Errors

As well as testing for correct answers, we can test that the behaviour is as expected when bad inputs are passed. This is even easier than our first test, requiring only one line of code.

```
test_that(
  "hypotenuse, with no inputs, throws an error",
  {
    expect_error(
```

```
    hypotenuse(),
    'argument "x" is missing, with no default'
  )
 }
)
```

    `expect_error` can take just one argument: a bit of code that you hope will throw an error. It is best practise to give it a second argument though. This is a regular expression to match the error message generated by the code.[2] By including the second argument, you avoid the possibility of a test erroneously passing by throwing the wrong error.

3.3.1   Exercise: Using `expect_error`

Write a test for `hypotenuse` when one or more inputs is a character vector.

3.4   Different Types of Expectation

`expect_equal` and `expect_error` are the most common *expectation* functions, but there are quite a few more that you ought to be aware of.

    Other common expectations are

- `expect_true`, `expect_false` and `expect_null`, which are shortcuts for checking those common return types.

- `expect_warning`, `expect_message` and `expect_output`, for testing feedback, which work like `expect_error`.

- `expect_identical`, a stricter check than `expect_equal`.

- `expect_equivalent`, a looser check than `expect_equal`, ignoring differences in attributes.

You may also occasionally come across these rarer expectations:

- `expect_length`, for checking the length of a vector or other object.

- `expect_lt` and `expect_gt`, for less than/greater than numeric inequalities.

- `expect_match`, for matching strings using regular expressions.

- `expect_type`, `expect_s3_class`, `expect_s4_class`, and the more general `expect_is`, for testing the type/class of variables.

---

[2]If the error is being generated by someone else's function, I usually cheat a bit by running the code to generate the error, then figure out the regular expression from that.

- `expect_named`, for testing the names of variables.

- `expect_silent`, for checking that there is no output.

- `expect_equal_to_reference` and `expect_output_file`, both perform `expect_equal` against a reference object stored in an `.rds`[3] or text file respectively.

- `expect_cpp_tests_pass`, for testing that C++ tests in a package pass.

> I sometimes struggle to remember the more obscure expectations. If you can't remember which expectation to use, you can get a complete list of them using `ls("package:testthat", pattern = "^expect_")`.

3.5 Testing Warnings and Messages

When code generates a warning, you need to test both the warning and the result of the code. This means that the test structure usually looks like a hybrid of the `expect_equal` and `expect_error` cases that we saw previously. The difference is that now the test needs to include at least two expectations.

Since `hypotenuse` doesn't throw any warnings, let's use the `min` function instead, which warns on a zero-length input. Notice that when we match the warning message, the second argument to `expect_warning` is a regular expression, so the hyphen needs to be escaped.

```
test_that(
  "min, with a zero-length input, returns infinity
  with a warning",
  {
    expected <- Inf
    expect_warning(
      actual <- min(numeric()),
      "no non\\-missing arguments to min; returning Inf"
    )
    expect_equal(actual, expected)
  }
)
```

For messages and printed output, the test structure is the same; the only change is to swap `expect_warning` for `expect_message` or `expect_output`.

---

[3] `.rds` files are a binary format for storing R objects. "RDS" stands for "R Data Serialization", and they are created with the base-R function `saveRDS`.

3.5.1 Chaining Expectations Together with Pipes

In Section 2.1.1 we saw how to chain several assertions together using *magrittr*'s forward pipe operator, `%>%`. This can also be used to chain expectations together. For example, the previous test can be re-written in the following way.

```
test_that(
  "min, with a zero-length input, returns infinity
  with a warning",
  {
    expected <- Inf
    expect_warning(
      numeric() %>% min,
      "no non\\-missing arguments to min; returning Inf"
    ) %>%
      expect_equal(expected)
  }
)

# Error: Test failed: 'min, with a zero-length input, returns
#   infinity with a warning'
# * '_lhs' not equal to 'expected'.
# target is NULL, current is numeric
```

To understand this code, you need to know that if the function on the right-hand side of the pipe only takes a single argument, you can just type the name of the function. If it takes more than one argument, then you need to write it as a function call with parentheses, passing all the arguments after the first one (which is automatically passed by the pipe).

3.5.2 Exercise: Using `expect_output`

Most `print` functions in R print a variable to the console (obviously), and invisibly return their first input argument. Write a test for the output and the return value of `print(1:5)`.

3.5.3 Testing for Lack of Output

The expectations for checking for conditions (the collective name for errors, warnings, and messages) and output also allow you to check for the opposite situation. If you pass `regexp = NA`, it implies that the condition should not occur.

The following example modifies the previous test of the minimum-value-of-nothing, to request that there is no warning, making it fail.

```
test_that(
  "min, with a zero-length input, returns infinity
  with no warning",
  {
    expected <- Inf
    expect_warning(
      actual <- min(numeric()),
      NA          # this line is different
    )
    expect_equal(actual, expected)
  }
)

# Error: Test failed: 'min, with a zero-length input, returns
#   infinity with no warning'
# * actual <- min(numeric()) showed 1 warning.
# *   no non-missing arguments to min; returning Inf
```

`expect_silent` provides a shortcut to check for no output, no messages, and no warnings.

3.6 Varying the Strictness of Expectations

`expect_identical` provides a stricter level of checking than `expect_equal`. You can use it for checking values that can be computed exactly, such as character vectors.

```
test_that(
  "the 1st, 5th, 9th, 15th and 21st letters are vowels",
  {
    expected <- c("a", "e", "i", "o", "u")
    actual <- letters[c(1, 5, 9, 15, 21)]
    expect_identical(actual, expected)
  }
)
```

For numeric values, the test is usually too strict due to floating-point rounding errors. For example, the following test will work with `expect_equal`, but fails with `expect_identical`.

```
test_that(
  "the square of the hypotenuse of 1, 1 is 2.",
  {
    expected <- 2
```

```
    actual <- hypotenuse(1, 1) ^ 2
    expect_identical(actual, expected)
  }
)
```

```
# Error: Test failed: 'the square of the hypotenuse of 1, 1 is 2.'
# * 'actual' not identical to 'expected'.
# Objects equal but not identical
```

In the opposite direction, **expect_equivalent** is less strict than **expect_equal**, in that it doesn't compare the attributes of the actual and expected values. This can be useful if your result has a name or a class, for example. In this case it is clearer to use multiple expectations to check the value and the attributes separately.

In the following example, **hypotenuse** will take the names of the first input, and include them in the output. We want to test both the return value and the names; using **expect_equivalent** allows us to ignore the names in the first expectation. In the next example, notice the use of **expect_named** as a shortcut for **expect_identical(names(actual), ...)**.

```
test_that(
  "hypotenuse returns input names in the output",
  {
    x <- c(a = 3)
    y <- 4
    expected <- 5
    actual <- hypotenuse(x, y)
    expect_equivalent(actual, expected)
    expect_named(actual, names(x))
  }
)
```

Most of the time, **expect_equal** is perfectly suitable; it should be your default choice.

3.7 Providing Additional Information on Failure

In the event that a test fails, *testthat* does reasonably well at explaining what the problem is. Sometimes however, a little more information can be informative to discover what went wrong, so *testthat* allows you to provide additional information in the failure message.

Consider this test that doesn't work.

```
test_that(
  "hypotenuse, with a NULL input, returns NULL",
  {
    expect_null(hypotenuse(3, NULL))
  }
)
```

```
# Error: Test failed: 'hypotenuse, with a NULL input, returns NULL'
# * hypotenuse(3, NULL) is not null.
```

On the face of it, the test seemed reasonable: put NULL into the function and get NULL out again. The test failure message correctly tells us that hypotenuse(3, NULL) isn't NULL, but it leaves us hanging with the question: what did the function return? By customising the info argument, we can find out straight away. In the next chunk of code, the base-R function deparse turns a value into the code that would generate that value.

```
test_that(
  "hypotenuse, with a NULL input, returns NULL",
  {
    actual <- hypotenuse(3, NULL)
    info <- paste("hypotenuse(3, NULL) =", deparse(actual))
    expect_null(actual, info = info)
  }
)
```

```
# Error: Test failed: 'hypotenuse, with a NULL input, returns NULL'
# * 'actual' is not null.
# hypotenuse(3, NULL) = numeric(0)
```

With additional information, we can quickly see that hypotenuse returned numeric(0), that is, a zero-length numeric vector.

Another use case for the info argument is running tests in loops: you can add additional information about which iteration of the loop caused the problem, for example.

3.8 Case Study: Calculating Square Roots

In this section we'll see how a testing workflow involves switching back and forth from writing tests and writing regular code as we find and fix issues.

The following function calculates square roots using the ancient Babylonian method.[4]

---

[4]Square root algorithms have moved on in the last few thousand years, so stick to the built-in sqrt function for real-world usage.

```
square_root <- function(x, tol = 1e-6)
{
  S <- x
  x <- log2(x) ^ 2
  repeat{
    x <- 0.5 * (x + (S / x))
    err <- x ^ 2 - S
    if(abs(err) < tol)
    {
      break
    }
  }
  x
}
```

To see if the function works correctly, let's start with some easy tests – giving it a positive number. We can use `base::sqrt` to compute the expected value since we know it gives the correct answer.

```
test_that(
  "square_root, with input 1024, returns 32",
  {
    expected <- 32
    actual <- square_root(1024)
    expect_equal(actual, expected)
  }
)
```

Modify the code to test **square_root** against another positive number. Does it work?[5]

The test passes, but how about a negative number input? The base-R `sqrt` function returns NaN with a warning, so our function should do the same.

```
test_that(
  "square_root, with a negative input, returns NaN with a warning",
  {
    expected <- NaN
    expect_warning(
      actual <- square_root(-1),
      "Negative inputs are not supported; returning NaN."
    )
    expect_equal(actual, expected)
  }
)
```

---

[5]You may discover that the algorithm is very slow for large inputs.

```
# Error: Test failed: 'square_root, with a negative input, returns
#    NaN with a warning'
# * missing value where TRUE/FALSE needed
# 1: expect_warning(actual <- square_root(-1), "Negative inputs are
#    not supported; returning NaN.") at <text>:5
# 2: capture_warnings(object)
# 3: withCallingHandlers(code, warning = function(condition) {
#          out$push(condition)
#          invokeRestart("muffleWarning")
#    })
# 4: square_root(-1)
```

Rather than warning about the negative input, `square_root` threw an error. The call stack from the error message is a bit of a monster, but can we see what went wrong? In this case, `log2(-1)` returns `NaN`, which feeds through to `err` being `NaN`, which then causes an error when it is used in the `if` condition. Now we can go back to our function and update it to handle the negative numbers more gracefully. Notice the use of `is_negative` from *assertive.numbers*.

```
square_root_v2 <- function(x, tol = 1e-6)
{
  if(is_negative(x))
  {
    warning("Negative inputs are not supported; returning NaN.")
    return(NaN)
  }
  S <- x
  x <- log2(x) ^ 2
  repeat{
    x <- 0.5 * (x + (S / x))
    err <- x ^ 2 - S
    if(abs(err) < tol)
    {
      break
    }
  }
  x
}
```

Now we need to re-run both the previous tests to see if our fix worked.

```
test_that(
  "square_root_v2, with input 1024, returns 32",
  {
    expected <- 32
    actual <- square_root_v2(1024)
    expect_equal(actual, expected)
```

```
    }
)
test_that(
  "square_root_v2, with a negative input, returns NaN with a
  warning",
  {
    expected <- NaN
    expect_warning(
      actual <- square_root_v2(-1),
      "Negative inputs are not supported; returning NaN."
    )
    expect_equal(actual, expected)
  }
)
```

This time, the lack of output shows that the tests passed. The issue with the `if` block points us to further problems that we might have, advising us on what to test next. Since an `if` condition must be either `TRUE` or `FALSE`, we might expect to have problems when x doesn't have length one.

```
test_that(
  "square_root_v2, with a zero-length numeric input,
  returns a zero-length numeric",
  {
    expected <- numeric()
    actual <- square_root_v2(numeric())
    expect_equal(actual, expected)
  }
)
```

```
# Error: Test failed: 'square_root_v2, with a zero-length numeric
#   input, returns a zero-length numeric'
# * argument is of length zero
# 1: square_root_v2(numeric()) at <text>:6
```

As expected, this fails on the `if` condition.

> When testing for zero-length inputs, it's quite common that you need two tests – a zero-length vector, and `NULL`. Write a test for a `NULL` input to **square_root_v2**.[6]

We can update our square root function yet again to deal with zero-length inputs.

---

[6]Hint: The expected value should be `numeric()`, not `NULL`.

```
square_root_v3 <- function(x, tol = 1e-6)
{
  if(is_empty(x))
  {
    return(numeric())
  }
  if(is_negative(x))
  {
    warning("Negative inputs are not supported; returning NaN.")
    return(NaN)
  }
  S <- x
  x <- log2(x) ^ 2
  repeat{
    x <- 0.5 * (x + (S / x))
    err <- x ^ 2 - S
    if(abs(err) < tol)
    {
      break
    }
  }
  x
}
```

Again we need to re-run the previous tests to make sure that they pass.

```
test_that(
  "square_root_v3, with input 1024, returns 32",
  {
    expected <- 32
    actual <- square_root_v3(1024)
    expect_equal(actual, expected)
  }
)
test_that(
  "square_root_v3, with a negative input, returns NaN
  with a warning",
  {
    expected <- NaN
    expect_warning(
      actual <- square_root_v3(-1),
      "Negative inputs are not supported; returning NaN."
    )
    expect_equal(actual, expected)
  }
)
test_that(
  "square_root_v3, with a zero-length numeric input,
  returns a zero-length numeric",
```

```
{
  expected <- numeric()
  actual <- square_root_v3(numeric())
  expect_equal(actual, expected)
}
)
```

You may have noticed a lot of repetition in this case study: tests need to be run many times! How to manage this repetition (without getting bored or going crazy) will be discussed in Section 3.11.

3.8.1   Exercise: Find More Tests for `square_root`

So far, we have tested positive numbers, negative numbers, and zero-length inputs to `square_root`. Make a list of as many other (useful) tests you can think of for the function.[7]

3.9   Other Expectations

There are a few more expectation functions that are more rarely seen. The structure of these tests is very similar to those that you've seen already – the purpose of the section is to give you some inspiration as to when you might actually want to use these tests.

`expect_type` tests that an object has a particular type, as determined by the base R function `typeof`. This can come in handy when you are using functions like `sapply` that can return results in different forms depending upon the dimensions of the input.

```
test_that(
  "log+sapply returns a (double) numeric vector when
    the input has length > 0",
  {
    actual <- sapply(1:5, log)
    expect_type(actual, "double")
  }
)

test_that(
  "log+sapply returns a list when the input has length 0",
  {
    actual <- sapply(numeric(), log)
    expect_type(actual, "list")
```

---

[7]If you are feeling enthusiastic, you can write those tests too.

```
  }
)
```

expect_s3_class and expect_s4_class test that an object inherits from a specific S3 or S4 class. For example, you would use expect_s3_class to test that the results of a linear regression have class lm. expect_is is an older, more general expectation that covers the three previous cases. For best practise, you should use one of the more specific functions.

expect_silent tests that your code has no output, messages, or warnings. This can be useful for testing a verbose argument to a function, where verbose = TRUE means "provide lots of feedback on progress," and verbose = FALSE means "keep quiet." The fread function in Matt Dowle and Arun Srinivasan's *data.table* package is very chatty when you set verbose = TRUE.

```
library(data.table)
test_that(
  "fread is silent when verbose = FALSE",
  {
    expect_silent(fread("x\ty\n1\ta", verbose = FALSE))
  }
)
test_that(
  "fread produces output when verbose = TRUE",
  {
    expect_output(fread("x\ty\n1\ta", verbose = TRUE), "")
  }
)
```

expect_length is used when you don't care about the specific values in a vector, but you want to know that you have a certain number of values. A typical use case would be checking residuals in a model, or checking results that involve random numbers. Since this is quite a weak expectation, it is common to use it in conjunction with further expectations in a test.

```
test_that(
  "rnorm gives a specified no. of results",
  {
    n <- 100
    actual <- rnorm(n)
    expect_length(actual, n)
    # + further expectations, e.g.
    expect_equal(mean(actual), 0, tol = 4 / sqrt(n))
  }
)
```

expect_lt, expect_gt, expect_lte, and expect_gte test for inequalities. Usually you want to test numbers using expect_equal, but if the answer if difficult to determine, sometimes it is easier just to test that they fall in a range. For example, you can test that a probability is greater than zero and less than one, without needing to bother calculating its exact value. Note that these expectations aren't vectorized; you have to test each value individually.

```
test_that(
  "pnorm always returns a probability",
  {
    actual <- pnorm(c(-Inf, 0, Inf))
    for(a in actual)
    {
      expect_gte(a, 0)
      expect_lte(a, 1)
    }
  }
)
```

expect_match tests that a string matches a particular regular expression. This can be useful for checking that character data is in a suitable format, without having to test for specific values. In the following example, we want to test that dates are in "yyyy-mm-dd" format, without caring what the actual date is. The datetime function in my *rebus.datetimes* [5] package is used to generate a regular expression that matches a valid date in this format.

```
library(rebus.datetimes)
test_that(
  "The data is in a yyyy-mm-dd format",
  {
    actual <- format(Sys.Date(), "%d-%m-%Y")
    regexp <- datetime("%Y-%m-%d")
    expect_match(actual, regexp)
  }
)

# Error: Test failed: 'The data is in a yyyy-mm-dd format'
# * 'actual' does not match
#   "(?:[0-9]{4}-(?:0[1-9]|1[0-2])-(?:0[1-9]|[12][0-9]|3[01]))".
# Actual value: "18-12-2016"
```

expect_equal_to_reference calls expect_equal against an object stored in an RDS file. It will be discussed in Section 6.2, Testing Complex Objects.

expect_cpp_tests_pass tests that C++ tests pass. It will be discussed in Section 6.4, Testing Rcpp Code.

3.9.1   Exercise: Testing the Return Type of Replicates

Write tests for the return type from a call to the base R function `replicate`, which runs a chunk of code repeatedly.

Hint: keep the chunk of code very simple.

3.10   Organising Tests Using Contexts

testthat doesn't place any restrictions on how many tests you have in a file. At one extreme, you could have one test per file; at the other, you could have all your tests in a single file. I find each extreme to cause problems with keeping track of where each test is – a middle ground where you have one test file per R file is typically easier to work with.

If you do end up with many tests in a file, you can break it into sections using *contexts*, via the function of the same name. This feature helps you remember what your tests are for, and helps keep them in a sensible order. When you come to run the tests, as we'll see in the next section, contexts also help you locate failing tests.

A typical use case is

```
context("Tests related to function1")
test_that("a test on function1", {whatever})
test_that("another test on function1", {whatever})

context("Tests related to function2")
test_that("a test on function2", {whatever})
test_that("another test on function2", {whatever})
```

3.11   Running Your Tests

As you've seen, a testing workflow involves running the same tests over and over several times. Each time you change a function, you need to re-run all the tests on it to make sure you haven't introduced new bugs. If you aren't careful, this can become really tedious.

Fortunately, *testthat* provides several functions that make it easy to repeatedly run your tests without going crazy. The most important of these is `test_check`, which we'll look at in Chapter 5, Integrating Testing into Packages. Outside of a package development context, the easiest way to run your tests is to use either `test_file` or `test_dir`, which run all the tests in a file or directory, respectively.

The following chunk of code contains three tests that we'll assume are in a
file named `test-file.R`. The first tests passes, the second test fails, and the
third test throws an error.

```
context("Some example tests")
test_that(
  "one is greater than zero",
  {
    expect_true(1 > 0)
  }
)
test_that(
  "one is less than zero",
  {
    expect_true(1 < 0)
  }
)
test_that(
  "this code runs",
  {
    stop("Oh no! An error!")
  }
)
```

To run the test in the file, we could just source this file. However, that
simply throws an error whenever each test fails, and if you have lots of tests
it can become hard to track which (and how many) tests failed on a given
run. `test_file` is more user friendly, giving you an easier to read summary
of what went wrong.

```
test_file("test-file.R")
```

```
# Some example tests: .12
#
# Failed ----------------------------------------------------------
# 1. Failure: one is less than zero (@test-file.R#11) ------------
# 1 < 0 isn't true.
#
#
# 2. Error: this code runs (@test-file.R#17) --------------------
# Oh no! An error!
# 1: stop("Oh no! An error!") at test-file.R:17
#
# DONE ============================================================
```

The output requires some explanation. The first line of the output tells you which tests passed (and the context of those tests, if you provided it) and which failed, in a concise manner. Each successful test is printed as a period. Tests that fail or throw an error are numbered: 1 to 9, then a to z, then A to Z; after that all failing tests are labelled F.

Subsequent lines of output contain the details of all the failures.[8] The output tells you whether an expectation wasn't met or the test threw an error, where the problem occurred in the file, and the output from the expectation or a stack trace of where the error occurred.

If you want to test more than one file, you can call test_dir to loop over all R files in a directory that start with the name "test".

> Try running the tests from the square root case study using test_file.

3.12 Customizing How Test Results Are Reported

If you don't like the output from test_file,[9] *testthat* provides a variety of output formats using *reporters*.

The minimal reporter displays the most concise output, merely displaying a dot for a pass, an "F" for a failure, or an "E" for an error.

```
test_file("test-file.R", reporter = "minimal")
```

```
# .FE
```

This sort of output is very useful when you are stuck trying to get that last tricky test or two to work. In this situation, you don't want to be bothered too much with lots of output as you tweak and rerun the test repeatedly – you just want to know whether the test finally worked or not.

The default reporter for test_file and test_dir is known as the summary reporter; it provides the output you saw in the previous section. You should use this one most of the time.

The stop reporter stops running tests the first time a test fails or an error occurs. This is useful for quick, interactive testing – it is the output you saw when the tests were run directly earlier in this chapter.

The silent reporter doesn't display any output, but invisibly returns a

---

[8]You actually only get the details of the first 61 labelled failures, though if you have more failing tests than this in a real-world situation, you probably need to rethink your testing workflow.

[9]If you don't like the output because it shows you that your tests are failing, that's a different problem.

data frame containing details of the errors. This allows you to programmatically inspect the errors, which can be useful in large projects where you need to filter the output to locate the most important tests to work on.

The `rstudio` reporter displays one line of output for each test failure or error. It's there in case the `summary` reporter is too verbose for your taste, and the `minimal` reporter is too brief.

The `tap` reporter provides output in the "Test Anything Protocol" standard format. Since this is a languge-agnostic format, it is useful for projects using multiple programming languages.

The `teamcity` reporter provides output in a suitable format for the Confluence TeamCity continuous integration platform. If you use TeamCity, then use this reporter; otherwise don't.

The `check` reporter is for use when building packages, which we'll look at in Chapter 5. It limits output to 13 lines, since that is all you can see when checking packages.

3.13   Alternatives

The main alternative to *testthat* is the *RUnit* package. This is based upon the *xUnit* framework, which makes the syntax very familiar if you have used Java's *JUnit* or Python's *unittest* testing frameworks. On the downside, it hasn't been developed in several years now, and is missing many key features that *testthat* provides. For example, there are no facilities for testing warnings, messages, or output. It also doesn't cache tests, which makes it much slower for large projects.

You can automatically convert tests generated using *RUnit* to the *testthat* syntax with the *runittotestthat* [6] package.

Brodie Gaslam's *unitizer* [?] is a new alternative that lets you interactively create tests. It's still ready for real-world use as of writing, but shows some promise.

3.14   Summary

In this chapter, you have

- Seen how to write unit tests for values, errors, warnings, and messages, using the *testthat* package.

- Experienced a development-time testing workflow.

- Seen how to run tests and control the display of the output.

4

Writing Easily Maintainable and Testable Code

This chapter describes some heuristics and techniques to help you write more easily maintainable and testable code. These ideas are commonplace for computer scientists and software engineers, but in the R community – where users are more likely to have a data analysis or statistics background – there isn't enough discussion on how to write good code.[1] If these ideas are new to you, read on. If not, skip to the package development chapters, and perhaps show this chapter to *that* colleague who writes the code that nobody wants to read.[2]

4.1   Don't Repeat Yourself

The principle of "Don't repeat yourself" has been nicely expressed as follows:

> Every piece of knowledge must have a single, unambiguous, authoritative representation within a system.
> http://c2.com/cgi/wiki?DontRepeatYourself

Apart from saving time by not writing things multiple times, there is a very good reason for avoiding duplication. Just to, ahem, repeat the point:

---

[1]My background was originally mathematics, and I spent several years writing awful code until I discovered these techniques.

[2]You could also recommend that they buy a copy because yay royalties.

> Duplicated code is bad code: anything that appears in two or more
> places in a program will eventually be wrong in at least one.
> http://software-carpentry.org/v4/essays/counting.html

That's the theory: if you make sure that all information appears in one
place in your code, then you can avoid bugs due to inconsistency. To see how
to put it into practice, let's take a look at an example.

4.1.1 Case Study: Drawing Lots of Plots

Consider this common workplace scenario. Near the end of a really big project
on the price of diamonds,[3] you show your diligently created report containing
dozens of insightful plots to your boss. Rather than praising you on your wonderful analysis, she explains that marketing have just changed the corporate
branding and you'll have to update all your plots from grey[4] to lime green in
order to follow corporate procedure.

You stare at your code, admiring the use of *magrittr*'s forward pipe operator to pass the data argument to the *ggplot2* [21] package's `ggplot` function
as you ponder how to change the colors.

```
library(ggplot2)
library(magrittr)
(scatter_price_vs_carat <- diamonds %>%
  ggplot(aes(carat, price)) +
  geom_point(color = "grey40")
)
```

---

[3]Of course you can substitute whatever is appropriate to your field of work. I chose
diamonds because the `diamonds` dataset in Hadley Wickham's *ggplot2* [21] package is rather
excellent for playing with.

[4]This was going to be blue but the author forgot to negotiate for color plots in the
contract.

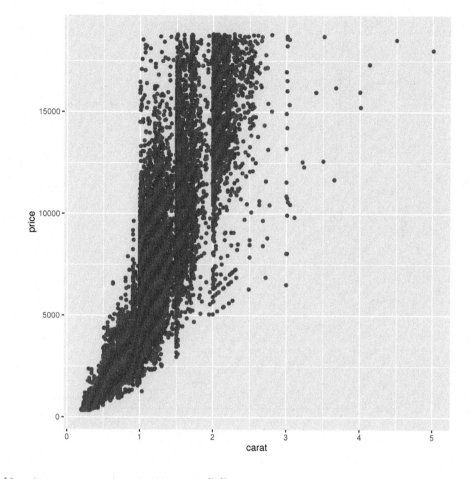

```
(density_cut_vs_price <- diamonds %>%
  ggplot(aes(price, color = cut)) +
  geom_density(color = "grey40")
)
```

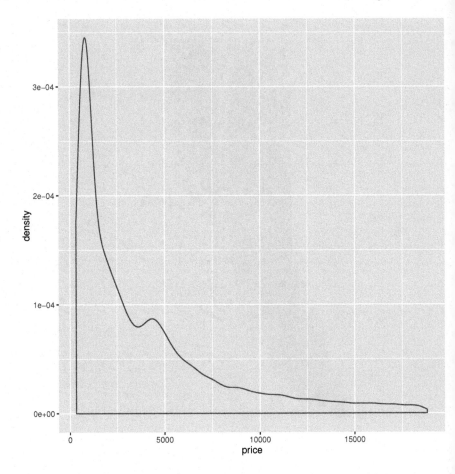

```
(bar_clarity_by_cut <- diamonds %>%
  ggplot(aes(clarity)) +
  geom_bar(fill = "grey40") +
  facet_wrap(~ cut)
)
```

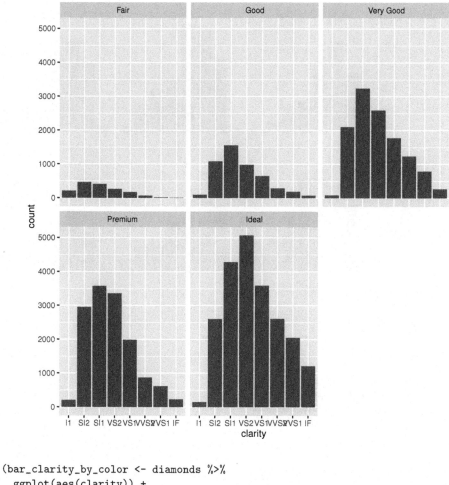

```
(bar_clarity_by_color <- diamonds %>%
  ggplot(aes(clarity)) +
  geom_bar(fill = "grey40") +
  facet_wrap(~ color)
)
```

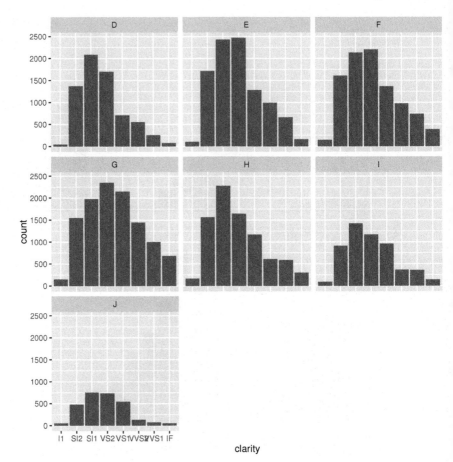

"grey40" is hard-coded into every single plot throughout your report. You could "find-and-replace-all," but that's risky, and you are too close to a deadline for a big disaster. Instead you decide to replace each instance of the color one at a time. Unfortunately, halfway through the update Sandra from Accounts phones you with an urgent query about your expenses last time you went to the useR! conference.[5] After the phone call you are distracted and accidentally miss a couple of plots, leaving your code inconsistent.

So how should you have written your code?

4.1.2   Idea 1: Use Variables Rather Than Hard-Coded Values

The simplest fix in this case is to replace all those hard-coded colors with a color variable. This way we only need to change the value in that assignment line.

---

[5]I'm sure the beer was absolutely a legitimate business expense.

```
col <- "grey40"          # <- change this to "limegreen"

(scatter_price_vs_carat <- diamonds %>%
  ggplot(aes(carat, price)) +
  geom_point(color = col) # <- color define by variable
)

(density_cut_vs_price <- diamonds %>%
  ggplot(aes(price, color = cut)) +
  geom_density(color = col)
)

(bar_clarity_by_cut <- diamonds %>%
  ggplot(aes(clarity)) +
  geom_bar(fill = col) +
  facet_wrap(~ cut)
)

(bar_clarity_by_color <- diamonds %>%
  ggplot(aes(clarity)) +
  geom_bar(fill = col) +
  facet_wrap(~ color)
)
```

This has solved our problem of only changing the color in one place, but there is still a lot of duplicated code, so there is still some scope for improvement.

4.1.3 Idea 2: For Values That You Want To Change Everywhere, Update Global Settings

If you have to use a corporate color scheme in every plot, it makes sense to set the default colors to this value. For just four plots, it isn't worth bothering with, but if you have hundreds of plots, then this will save you some time.

For the general case – outside of *ggplot2* – most global options can be changed using `options`. Here we need to use `update_geom_default` to change the default plot appearance.[6].

In the following code block, notice how we don't need to specify the color at all in the code for creating the plots; it has already been taken care of. This eliminates the duplicated instances of `color = col` and `fill = col` that we saw before.

```
col <- "grey40"   # <- change this to "limegreen"
```

---

[6]Note that you currently have to use British spellings of aesthetic name with `update_geom_default`

```
update_geom_defaults("point", list(colour = col))
update_geom_defaults("density", list(colour = col))
update_geom_defaults("bar", list(fill = col))

# Now we can drop the code specifying the colors
(scatter_price_vs_carat <- diamonds %>%
  ggplot(aes(carat, price)) +
  geom_point()
)

(density_cut_vs_price <- diamonds %>%
  ggplot(aes(price, color = cut)) +
  geom_density()
)

(bar_clarity_by_cut <- diamonds %>%
  ggplot(aes(clarity)) +
  geom_bar() +
  facet_wrap(~ cut)
)

(bar_clarity_by_color <- diamonds %>%
  ggplot(aes(clarity)) +
  geom_bar() +
  facet_wrap(~ color)
)
```

4.1.4   Idea 3: Wrap the Contents into a Function

The two bar plots share a lot of code, so one way of eliminating even more code is to write a function that can create each plot. For just two plots, it's only borderline worth the effort, but as you increase the number of plots, the more useful this technique is.

In the following example, we use a combination of **paste** and **as.formula** to convert a string containing the name of the variable to facet by into a formula suitable to be passed to *ggplot2*'s **facet_wrap**. This way, each bar plot can be created with a single function call.

```
barplot_diamond_clarity <- function(facet_var)
{
  facet_formula <- as.formula(paste("~", facet_var))
  diamonds %>%
  ggplot(aes(clarity)) +
    geom_bar() +
    facet_wrap(facet_formula)
}
```

```
bar_clarity_by_cut <- barplot_diamond_clarity("cut")
bar_clarity_by_color <- barplot_diamond_clarity("color")
```

Encapsulating code inside functions has another advantage: by putting the functions inside packages, you can reuse the code across multiple projects and share it with other people.

> Can you think of any other ways to reduce duplication in this code?

4.1.5 Exercise: Reducing Duplication

Have a look at the `ChickWeight` dataset on weights of chickens under different diets.[7] Use, `str`, `head`, `summary`, and any other functions you need to understand the dataset.

Here is some repetitive code to calculate the mean chicken weight for each diet.

```
mean(ChickWeight$weight[ChickWeight$Diet == 1])

# [1] 102.6455

mean(ChickWeight$weight[ChickWeight$Diet == 2])

# [1] 122.6167

mean(ChickWeight$weight[ChickWeight$Diet == 3])

# [1] 142.95

mean(ChickWeight$weight[ChickWeight$Diet == 4])

# [1] 135.2627
```

Rewrite the code in a way that eliminates the duplication.

(There are *lots* of ways of doing this in R; if you are feeling confident, see how many different solutions you can find.)

4.2 Keep It Simple, Stupid

There's a famous result from cognitive psychology dating back to the 1950s that very roughly says

---

[7]Not to be confused with `chickwts`; R has two built-in datasets related to chicken weights!

> On a good day, you can keep seven (give or take two) things at once
> in your working memory.
>
> Miller's Law, paraphrased

Since R is a high-level programming language where one line of code (again, very roughly) translates into one useful thought, I present the Cotton corollary:

> Most of your R functions should be seven lines or less.
>
> The Cotton Corollary

Of course, not every function can be squeezed into such a short space, at least not in any way that actually benefits you. If you are regularly writing R functions that you can't read on a single screen however, you are probably making life more complicated for yourself[8]. To hammer a point home, it's perfectly acceptable for a function to be one line long. These are usually the easiest kind of function to debug.

4.2.1   Simplifying Function Interfaces

As well as keeping functions short, the other side of keeping things simple is to make it easy for users to understand how to call your function. Making functions do more things makes their interface harder to use; it also means that you need more tests for them.

My *sig* [7] package, for working with function *signatures*,[9] has a `sig_report` function to find all the functions in an environment (or package) that are very long or have too many inputs. Some of the packages in the base-R distribution are surprisingly bad offenders. Here's a report on the *utils* package.

```
library(sig)
sig_report(pkg2env(utils))
```

---

[8]I once had to help a customer debug a 3000 line R function that they'd written. Please do not write functions this long yourself.

[9]In the context of R programming, *signature* just means the input arguments to a function. In other languages, it can also include the return value, but since you can't force that to be a particular type in R, there's no point in worrying about it.

```
# The environment contains 209 variables of which 206 are functions.
# Distribution of the number of input arguments to the functions:
#  0  1  2  3  4  5  6  7  8  9 10 11 12 13 14 16 25
#  7 43 38 28 22 21 12  5 15  5  2  2  2  1  1  1  1
# These functions have more than 10 input args:
# [1] example          help.search      rc.settings
# [4] read.DIF         read.fwf         read.table
# [7] update.packages write.table
# Distribution of the number of lines of the functions:
#           1         2       [3,4]      [5,8]      [9,16]     [17,32]
#           0        38          10         21          32          27
#     [33,64]   [65,128]   [129,256]  [257,512]
#          42        24          11          1
# These functions have more than 50 lines:
# [1]  ?                        .S3methods
# [3]  adist                    aspell
# [5]  assignInNamespace        available.packages
# [7]  bibentry                 browseEnv
# [9]  changedFiles             citation
# [11] citeNatbib               combn
# [13] create.post              data
# [15] de.restore               demo
# [17] download.packages        example
# [19] findLineNum              getAnywhere
# [21] help                     help.search
# [23] hsearch_db               localeToCharset
# [25] make.packages.html       makeRweaveLatexCodeRunner
# [27] package.skeleton         packageDescription
# [29] packageStatus            person
# [31] promptData               read.DIF
# [33] read.fwf                 read.table
# [35] RShowDoc                 RweaveLatexWritedoc
# [37] select.list              sessionInfo
# [39] summaryRprof             Sweave
# [41] tar                      txtProgressBar
# [43] untar                    unzip
# [45] update.packages          write.table
```

The report shows 8 functions with more than 10 input arguments – in fairness several of these are reading and writing functions which do require a lot of customizability – and 46 functions with more than 50 lines.

Try running a **sig_report** on the *Hmisc* package, the gold standard in monstrously big and complicated functions. For contrast, try running a report on *assertive* and *testthat* too.

Now you can identify which functions you need to simplify, but how do you go about making them simpler?

There's a rather obvious piece of advice ("You Ain't Gonna Need It") that states that you shouldn't write code for features that you don't need.

Always implement things when you *actually* need them, never when
you just *foresee* that you need them.
http://c2.com/cgi/wiki?YouArentGonnaNeedIt

Assuming that you've only implemented what you need to implement, and it's still a bit tricky, how can you make it easier to use? Let's take a look at a few complicated functions, and see how they manage complexity.

4.2.2   Idea 1: Pass Arguments for Advanced Functionality to Another Function

In a function signature, not all the input arguments have equal importance. In many, dare I say *most*, functions, the first one or two arguments are the most important, and the rest are there for advanced usage and tweaking. This means that, even if you have a function signature of twenty arguments, it often just means that the function looks harder to use than it actually is. One way to help your users figure out which arguments can be safely considered "advanced usage only," and ignored while they learn how to use the function, is to outsource them to a helper function that processes input arguments. The *grid* package makes heavy use of this technique.

Let's use the `sig` function, from the package of the same name, to take a look at the signature for one of the *grid* functions.

```
library(grid)
sig(pointsGrob)
```

```
# pointsGrob <- function(x = stats::runif(10), y =
#    stats::runif(10), pch = 1, size = unit(1, "char"),
#    default.units = "native", name = NULL, gp = gpar(), vp = NULL)
```

In the function signature for many of the functions in *grid*, you see `gp = gpar()`. `gpar` just checks that its inputs are sensible, then returns them as a list:

```
gpar(col = "red", cex = 3)

# $col
# [1] "red"
#
# $cex
# [1] 3
```

You can use them as follows:

```
pointsGrob(1:10, 10:1, gp = gpar(col = "red", cex = 3))

# points[GRID.points.673]
```

This allows all the argument checking to be written once (in **gpar**) rather than appearing in every *grob*[10] function.

4.2.3   Exercise: Outsourcing Argument Checking

Max Kuhn's *caret* [11] package contains tools for training classification and regression models. Use the **sig** function to see the signature for the **train.default** function.

One of the arguments is **trControl = trainControl()**, which is similar to *grid*'s **gpar**, in that it accepts advanced training arguments, checks that they are suitable, then returns them as a list to be used by **train.default**.

Compare these uses of **trainControl**:

```
library(caret)
trainControl()
trainControl(method = "adaptive_LGOCV", classProbs = TRUE)
trainControl(returnResamp = "any")
```

Try passing bad arguments to **trainControl** to see what errors you can make it throw. Does it catch all possible bad inputs?

What extra checks could you add to **trainControl** to improve it?

Write some unit tests to make sure that your updated function really works as you think it should.

---

[10]*grob* is short for "graphics object." Plots created using grid graphics, such as *ggplot2* or *lattice* plots, consists of lots of grobs.

4.2.4 Idea 2: Having Wrapper Functions for Specific Use Cases

The `read.table` function in the *utils* package is pretty complicated. It needs to cope with a lot of variation in file formats, so it necessarily has a lot of arguments.

```
sig(read.table)
```

```
# read.table <- function(file, header = FALSE, sep = "", quote =
#    """, dec = ".", numerals = c("allow.loss", "warn.loss",
#    "no.loss"), row.names, col.names, as.is = !stringsAsFactors,
#    na.strings = "NA", colClasses = NA, nrows = -1, skip = 0,
#    check.names = TRUE, fill = !blank.lines.skip, strip.white =
#    FALSE, blank.lines.skip = TRUE, comment.char = "#",
#    allowEscapes = FALSE, flush = FALSE, stringsAsFactors =
#    default.stringsAsFactors(), fileEncoding = "", encoding =
#    "unknown", text, skipNul = FALSE)
```

In order to reduce complexity for users, *utils* also provides four wrapper functions for the most common cases: `read.csv` and `read.delim` for reading comma and tab delimited files, respectively, and their variants `read.csv2` and `read.delim2` which use European-style commas for decimal places. Each of the wrapper functions provides default arguments suitable for their specific use-case, meaning that the user doesn't have to bother setting them. For example, take a look at the signature for `read.csv`.

```
sig(read.csv)
```

```
# read.csv <- function(file, header = TRUE, sep = ",", quote =
#    """, dec = ".", fill = TRUE, comment.char = "", ...)
```

4.2.5 Exercise: Wrappers for Formatting Currency

`formatC` provides C-style format specifications for numbers. It has a very flexible but complicated interface. Take a look at it using the `sig` function.

Write a wrapper function for the specific use case of printing US currency. That is, the function should accept a numeric vector and return a character vector with the number given in fixed format[11] to two decimal places, and prefixed by a dollar sign.

---

[11]As opposed to scientific format.

4.2.6 Idea 3: Auto-Guessing Defaults

Another approach to solving the `read.table` complexity issue is found in the `fread` function of the *data.table* package. While the focus is mainly on the speed of reading files, it is interesting to see how the authors have tried to improve on speed of usage. The function signature isn't much simpler:

```
library(data.table)
sig(fread)
```

```
# fread <- function(input = "", sep = "auto", sep2 = "auto", nrows
#    = -1, header = "auto", na.strings = "NA", file,
#    stringsAsFactors = FALSE, verbose =
#    getOption("datatable.verbose"), autostart = 1, skip = 0,
#    select = NULL, drop = NULL, colClasses = NULL, integer64 =
#    getOption("datatable.integer64"), dec = if (sep != ".") "."
#    else ",", col.names, check.names = FALSE, encoding =
#    "unknown", quote = "\"", strip.white = TRUE, fill = FALSE,
#    blank.lines.skip = FALSE, key = NULL, showProgress =
#    getOption("datatable.showProgress"), data.table =
#    getOption("datatable.fread.datatable"))
```

`fread` does, however, make its usage easier (and hence faster) than `read.table`. Its trick is to do a lot of automated guesswork as to what arguments to use. For example, with `read.table` you have to inspect your file, maybe with a text editor, to see whether the separator is a comma or a tab or something else. `fread` saves you this trouble by doing smart guesswork to automatically determine what the separator is. So, while you can manually set the `sep` argument, you almost never have to. Likewise, `read.table` needs you to state whether or not your data has a header row, but `fread` is smart enough to correctly guess this. Similarly, there is smarter guessing of column class types and the number of rows.

While the effect of automatically guessing individual arguments is only a small timesaver, the cumulative effect of having most arguments dealt with means that you can devote much less brainpower to this task.

4.2.7 Exercise: Providing Better Defaults for `write.csv`

`write.csv` in the *utils* package has some slightly annoying[12] defaults. Write a wrapper for this function that defaults to not writing row names to the file, and that writes missing values as an empty cell.

For bonus points, autogenerate the default file name from the input variable as well.

---

[12] In the author's humble opinion.

4.2.8 Idea 4: Split Functionality into Many Functions

Unix has a well-known philosophical standpoint that programs should

---

Do One Thing and Do It Well

Doug McIlroy, 1978

http://www.faqs.org/docs/artu/ch01s06.html

In the *plyr* [22] package, `ddply` is very powerful, but the interface takes a while to learn. This next example, gets the mean weight of chickens by feed group.

```
library(plyr)
ddply(chickwts, .(feed), summarize, MeanWeight = mean(weight))

#          feed MeanWeight
# 1      casein   323.5833
# 2 horsebean   160.2000
# 3    linseed   218.7500
# 4   meatmeal   276.9091
# 5    soybean   246.4286
# 6 sunflower   328.9167
```

This is already quite complex, but `ddply` also allows you to do other tasks, like adding additional columns to the data or calculating summary stats on multiple columns simultaneously. The *dplyr* approach, which (mostly) replaces *plyr*, is to split the functionality into several different functions. In this next example, `%>%` is the pipe operator, which passes the result of an expression into the first argument of the next function, allowing you to join multiple commands together.

```
library(dplyr)
chickwts %>%
  group_by(feed) %>%
  summarize(MeanWeight = mean(weight))

# # A tibble: 6   2
#          feed MeanWeight
#        <fctr>      <dbl>
# 1      casein   323.5833
```

```
# 2 horsebean   160.2000
# 3   linseed   218.7500
# 4  meatmeal   276.9091
# 5   soybean   246.4286
# 6 sunflower   328.9167
```

4.2.9   Exercise: Decomposing the `quantile` Function

The `quantile.default` function, internal to the *stats* package, is a great example of a function that is an awful incomprehensible mess because it contains too many ideas in a single place. Take a look at the code for `stats:::quantile.default` and describe how you could decompose it in order to make it more readable. (You don't need to provide an implementation.[13])

4.2.10   Cyclomatic Complexity

As well as keeping functions short, another way of keeping it simple is to reduce the number of possible paths through your code. Whenever you include `if` or `switch` statements, or loops, then you increase the number of possible ways that the function can run, and you make it harder to reason about what your code is doing. This measure of how many ways a function can run is known as *cyclomatic complexity*.

The cyclomatic complexity number (CCN) is a measure of how many paths there are through a method. It serves as a rough measure of code complexity and as a count of the minimum number of test cases that are required to achieve full code-coverage of the method.

R. K. Cole
http://www.rkcole.com/articles/other/CodeMetrics-CCN.html

... or alternatively:

Cyclomatic complexity measures the number of linearly independent paths through the method, which is determined by the number and complexity of conditional branches. A low cyclomatic complexity generally indicates a method that is easy to understand, test, and

---

[13]Of course, if you are feeling particularly enthusiastic then feel free to have a go at an implementation.

maintain.

Microsoft

https://msdn.microsoft.com/en-us/library/ms182212.aspx

In the simplest case, we can write a function where there is only one path to take. The cyclomatic complexity is calculated using Gabor Csardi's *cyclocomp* [9] package.

```
library(cyclocomp)
cyclo_single_path <- function()
{
  message("Hello World!")
}
cyclocomp(cyclo_single_path)

# [1] 1
```

Let's increase the complexity by including an `if` statement.

```
cyclo_if <- function(condition)
{
  if(condition)
  {
    message("Hello World!")
  }
}
cyclocomp(cyclo_if)

# [1] 2
```

`cyclocomp` calculates this as having a cyclomatic complexity of two: you can pass either `TRUE` or `FALSE` to the condition, and it will change the behaviour. For most programming languages, this is absolutely correct. However, in R, you can also pass `NA`, which throws an error, so the cyclomatic complexity is arguably three. (You can also pass numbers or strings or logical vectors with length greater than one, but these cases always resolve back to one of the three cases of `TRUE`/`FALSE`/`NA`.) The upshot of this is that cyclomatic complexity isn't very well defined in R, so don't worry about the count; the important point is that having lots of branching within a function can make things harder to understand.

In the following example, using nested `if` statements makes the code fairly awful to read, a fact that is reflected in its bigger cyclomatic complexity score.

```
cyclo_nested_if <- function(date)
{
  if(some_condition)
  {
    if(some_other_condition)
    {
      if(a_third_condition)
      {
        all_three_things
      } else
      {
        the_first_two_things
      }
    } else
    {
      just_the_first_thing
    }
  } else
  {
    not_the_first_thing
  }
}
cyclocomp(cyclo_nested_if)
```

As well as `if` and `else`, other language features that increase the cyclomatic complexity include loops (`for`, `while`, and `repeat`); and branching with `ifelse`, `switch`, and the logical `&&` and `||` operators.

4.2.11 How to Reduce Cyclomatic Complexity

1. Due to R's trooolean logic, dynamic typing, and lack of scalar types, there are a lot of "bad" cases, so it is good practise to check that you genuinely have a scalar logical value to pass to an `if` statement, as early as possible in your functions.

2. Nested `if` statements of loops quickly increase the cyclomatic complexity of the code. These are often an indication that some of the logic needs to be outsourced into a different function.

3. Early returns for edge cases can be very useful. For example, if you have a zero-length input, you might just want to return a zero-length output straight-away, without bothering with other calculations. This can save you worrying about special situations later in the function.

4. Refactoring the function into smaller functions reduces complexity until it doesn't.

4.2.12   Exercise: Calculating Leap Years

Here's a function that accepts a year (as a number) and returns TRUE or FALSE depending upon whether or not that year is a leap year. The logic is that leap years occur on years divisible by 4, excluding centuries, but including every fourth century. Recall that the %% operator means remainder after division.

```
is_leap_year <- function(year)
{
  if(year %% 4 == 0)
  {
    if(year %% 100 == 0)
    {
      if(year %% 400 == 0)
      {
        TRUE
      } else
      {
        FALSE
      }
    } else
    {
      TRUE
    }
  } else
  {
    FALSE
  }
}
```

The nested if statements make the code hard to understand; they also stop the code being vectorized (it only accepts one input number at a time). Rewrite the function to make it easier to read. For bonus points, vectorize it.

4.3   Summary

- Repeated code can lead to inconsistency bugs; try to avoid duplication.

- In general, functions should be short and have a simple interface. Use the *sig* package to identify problem functions.

- Code with lots of branching and loops has a high cyclomatic complexity and may need to be split into smaller components. Use the *cyclocomp* package to identify problem functions.

5

Integrating Testing into Packages

The success of R is due, in a large part, to the contributions of users.[1] This chapter tells you how to create a package, with tests, so you can contribute too.

5.1 How to Make an R Package

Making packages is mostly easier than you think: it's mostly just a case of putting your files in a specific, well-defined file structure and clicking the "build" button in your IDE. There is a bit of an art to it, however, particularly if you want to do something more advanced. A full reference for how to make packages is way beyond the scope of this book; we'll just look at the very basics here. If you want to know more, *Learning R* [4] has a chapter that takes you step-by-step through making your first package, and *R Packages* [25] by Hadley Wickham is a wonderful, thorough guide to the subject. The official documentation, *Writing R Extensions*, isn't quite as easy to read, but contains many explanations of all the technicalities of package creation, and details of what is allowed and what is forbidden. Of course, when things go wrong, programming question and answer site *Stack Overflow* [19], and the *R-package-devel* [20] mailing list are good sources of solutions.

5.1.1 Prerequisites

Under Linux and other Unix-based operating systems, you'll typically have all the necessary tools that you need to build packages. For Windows, you need to install *Rtools*. This is most easily done by using the `install.rtools` function in Tal Galili's *installr* [13] package, or you can download it from http://cran.r-project.org/bin/windows/Rtools.

```
library(installr)
install.rtools()
```

---

[1]Not to belittle the wonderful work by the R-Core Team, of course.

You also need the *devtools* [27] and *roxygen2* [28] packages, both by Hadley Wickham.

5.1.2   The Package Directory Structure

The name of the package directory is, by convention, the same as the name of the package. At the top level there are two compulsory files: `DESCRIPTION` and `NAMESPACE`. `DESCRIPTION` contains information like the title, description, and authors of the package, and requires some manual editing by you. `NAMESPACE` lists all the functions exported by the package (so that users can use them) and the functions imported from other packages. This file can be automatically generated by *roxygen2*.[2]

The names of the directories that are allowed in the top level are tightly specified, with two compulsory directories and a limited number of optional directories. The `R` directory contains your code, and `man` contains the help files.

Two other optional directories are important to us. `src` contains C, C++, or FORTRAN source code. We'll make use of this directory in the next chapter. `tests` contains our unit tests, which we'll make use of right away![3]

5.1.3   Including Tests in Your Package

What goes into this `tests` directory you may ask? We need two things for testing with *testthat*: a directory of R files containing our tests, and another R script to run them.

The R files containing the tests go in a directory named *testthat*.[4] Each of the R files should have a name beginning with "test-." With this naming convention, tests are automatically found and run when we come to check the package.

The script to run the tests is conventionally called `testthat.R` and has very specific contents, as follows.

```
library(testthat)
library(the_package_you_are_testing)

test_check("the_package_you_are_testing")
```

You can think of `test_check` as being the package equivalent of `test_dir` that we saw back in Chapter 3.

---

[2]There are three other optional files, but you don't need to worry about them now. `LICENSE` or `LICENCE` is used when you are using a non-standard license for the package. `NEWS` describes what changes you've made to the package. `INDEX` is an automatically generated file that describes the interesting objects in the package.

[3]Veteran package developers may notice that this has changed. Tests used to go in `inst/tests`.

[4]That is, `tests/testthat`, relative to the root of the package.

Beyond the `tests/testthat` directory and the `testthat.R` file, you also need to make sure that the *testthat* package is listed in the *Suggests* field of the `DESCRIPTION` file.[5]

The easiest way to add this file infrastructure is to call the `use_testthat` function in the *devtools* package.

```
library(devtools)

use_testthat("path/to/your/package")
```

5.2   Case Study: The *hypotenuser* Package

All that was a bit theoretical, so let's see how it works with a real example. We'll package the `hypotenuse` function that you saw in Chapter 3. Let's call it the *hypotenuser* package.[6]

> A copy of this package is available at
> https://bitbucket.org/richierocks/hypotenuser
> By viewing the commits (on the left-hand menu; more on this in a moment), you can see how the package changes at each step.

To create the basic files and directories for a package, we'll use `create` from the *devtools* package. This takes the name of the package and a `description` argument that describes what fields go into the `DESCRIPTION` file.

`create` creates the `DESCRIPTION` and `NAMESPACE` files and an R directory. If you set `rstudio = TRUE` (the default), then you also get some files for working with the RStudio IDE: a `.Rproj` project file,[7] a `.Rbuildignore` file that tells R to ignore RStudio-related files when building the package, and a `.gitignore` file that tells git not to track RStudio-related files.

If you prefer, you can manually create the package using your operating system's file explorer. Create a directory named `hypotenuser`, a subdirectory named R, and a text file named `DESCRIPTION`. (The other files are optional.) I recommend using `create` though, since it's very easy to reuse the creation script each time you create a new package, simply changing the Title, Description, URL, and BugTracker fields. For brevity, the output of the following code is omitted.

---

[5] *Suggests* is for package dependencies that only crop up in tests, examples, and vignettes.

[6] Naming things is hard. Feel free to choose a better name when you try out this code.

[7] These project files are also used by Microsoft's R Tools for Visual Studio, which lets you run R code in the Visual Studio IDE.

```
library(devtools)
descriptionDetails <- list(
  Title       = "Calculate Hypotenuses",
  Version     = "0.0-1",
  Author      = "Summer Squares [aut,cre]",
  Maintainer  = "Summer Squares <s@sq.com>",
  Description = "A hypotenuse fn, plus tests!",
  License     = "GPL-3",
  URL         = "https://package-homepage.com",
  BugReports  = "https://package-homepage.com/issues"
)
create("hypotenuser", description = descriptionDetails)

# Creating package 'hypotenuser' in
#   '/Users/richierocks/workspace/testingrcode/chapters'
# No DESCRIPTION found. Creating with values:
# * Creating 'hypotenuser.Rproj' from template.
# * Adding '.Rproj.user', '.Rhistory', '.RData' to ./.gitignore
```

The contents of the DESCRIPTION file are shown in Figure 5.1.

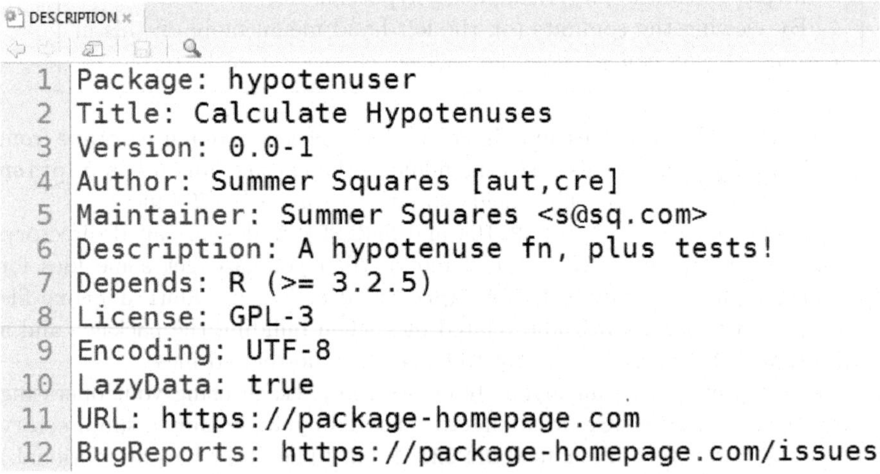

```
 1  Package: hypotenuser
 2  Title: Calculate Hypotenuses
 3  Version: 0.0-1
 4  Author: Summer Squares [aut,cre]
 5  Maintainer: Summer Squares <s@sq.com>
 6  Description: A hypotenuse fn, plus tests!
 7  Depends: R (>= 3.2.5)
 8  License: GPL-3
 9  Encoding: UTF-8
10  LazyData: true
11  URL: https://package-homepage.com
12  BugReports: https://package-homepage.com/issues
```

FIGURE 5.1
The contents of the DESCRIPTION file for the *hypotenuser* package.

Now that we have the basic file structure, we need to add some content: the hypotenuse function. We will save the following code to a file named hypotenuse.R. The lines beginning #' are *roxygen2* comments that will be used to generate the help page for the function.

```
#' Calculate hypotenuses
#'
#' Calculates the hypotenuse, using the obvious algorithm.
#' @param x A numeric vector.
#' @param y Another numeric vector.
#' @return The hypotenuse.
#' @export
hypotenuse <- function(x, y)
{
  sqrt(x ^ 2 + y ^ 2)
}
```

To add the directory for testing, we call **use_testthat**.

```
use_testthat("hypotenuser")

# * Adding testthat to Suggests
# * Creating 'tests/testthat'.
# * Creating 'tests/testthat.R' from template.
```

The final bit of content to add to the package is a file of tests, inside the **tests/testthat** directory. We'll save the following tests to a file named **test-hypotenuse.R**.

```
context('Testing the hypotenuse function')
test_that(
  'hypotenuse, with inputs 5 and 12, returns 13',
  {
    expected <- 13
    actual <- hypotenuse(5, 12)
    expect_equal(actual, expected)
  }
)
test_that(
  'hypotenuse, with inputs both 1e300, returns sqrt(2) * 1e300',
  {
    expected <- sqrt(2) * 1e300
    actual <- hypotenuse(1e300, 1e300)
    expect_equal(actual, expected)
  }
)
```

Now that the package contents are in place, to make it useable, we need to build it. You can do this through the user interface of your R IDE, or use build from the *devtools* package. Building shows a fair amount of output, not included here for brevity's sake.

```
build("hypotenuser")
```

5.3　Checking Packages

As well as building packages, R also has functionality for checking packages. It's good practise to run these checks regularly, to ensure that your package is in good shape. The list of checks is rather extensive; some of the more important checks include

- ensuring the DESCRIPTION is correctly formed.

- making sure that all the necessary dependency packages are declared in the DESCRIPTION and NAMESPACE files.

- that all the help pages are correctly formed (and that all the arguments in the functions they describe are mentioned).

- checking the R code for syntax errors.

- making sure that all the necessary files are present, and that there are no unknown files.

- running all the examples in the help pages, and making sure there are no errors.

- running all the tests, and making sure they all pass.

> Including examples in help pages, and running R's check functionality, provide a form of package testing *for free*. Make sure you use them!

As with building, the easiest way to check a package is via the user interface of your IDE, but you can also call check.[8] Again, the output isn't shown here for brevity.

```
check("hypotenuser")
```

---

[8]If you are a shell enthusiast, you can also call R CMD check hypotenuser from your operating system command line.

The Bioconductor repository has a package named *BiocCheck* [18]. This package performs additional checks, like ensuring that at least 80% of your help pages have a runnable example. Using it is compulsory if you want to submit your package to Bioconductor, but most of the checks apply equally well to other packages, so it's a good idea to use it for all your packages.

5.3.1 Exercise: Make a Package with Tests

Bundle the `square_root_v3` that we looked at in Chapter 3, along with at least one of the tests we wrote.

5.4 Using Version Control, Online Package Hosting, and Continuous Integration

As we've already discussed, the main point of development-time testing is to ensure that you haven't made any silly mistakes. As well as unit tests, there are some other tools to help you spot mistakes and to help you fix them.

5.4.1 Version Control with *git*

Version control software provides an "unlimited undo" feature for code. It is arguably the most important tool that you need if you want to write high-quality code.[9]

The idea is that, after you make some changes to your code (a new feature, or a bug fix perhaps), you *commit* them and *push* them to a repository. If other people are working on the same piece of software, they can *pull* your changes into their copy, thus sharing your work. The beauty is that each commit acts as a checkpoint. If you realize that you've made a mistake (and let's face it, mistakes happen), you can roll back to a previous commit, and pretend it never happened.

Currently, the most popular version control software is *git*, which you can download from

https://git-scm.com/download

git is a command line tool, but the RStudio and Visual Studio IDEs both come with git integration, so you can commit and push and pull with a graphical user interface. Architect, the Eclipse-based R IDE, doesn't ship with git integration, but it is available via the *egit* plugin.

A full tutorial on how to use git is beyond the scope of the book. There are many introductory guides; the one I find to be the most fun and least painful is on the Software Carpentry site:

---

[9]Or any other text document, like this book, for example.

https://swcarpentry.github.io/git-novice

5.4.2   Online Project Hosting

git on its own leaves how you share your code up to you. Consequently several websites have sprung up that host your code projects, acting as a git repository, with tools to make sharing code, tracking bugs, and collaborating very easy.

In the R community github (https://github.com) is the most popular, with Bitbucket (https://bitbucket.org) in second place. Both platforms have a similar feature list and a reasonably similar user interface. github makes it a little easier to provide a website to go with your project, and its larger user base means that you're slightly more likely to get other people helping you. Bitbucket, on the other hand, let's you have unlimited free private repositories, which allows you to restrict access to your packages until they are ready to be seen by the world (often important in a corporate setting).

If you don't have accounts to these services already, set them up now. They both have free plans available, and it only takes a few minutes to register.

There are several thousand R packages available on github and Bitbucket; it is quite common to have a development version of your R package on one of these, and the stable version on CRAN or Bioconductor. You may have come across these platforms from accessing these packages through *devtools*. You can install a package from these places using:

```
devtools::install_github("username/packagename") # or
devtools::install_bitbucket("username/packagename")
```

5.4.3   Continuous Integration Services

Although it's very easy to run R's package check (only a click or a keystroke in most IDEs), checking a package in all possible use situations takes considerably more effort. For example, there may be differences in behaviour between the current release of R and the development version. If you have any code that involves file paths or locales,[10] then R's behaviour can be different across operating systems. It's also possible to have unusual check results on your own machine just because you have some custom configuration set up.

What this entails is that it's a really great idea to check your package

---

[10]Location and culture-specific behaviour, including settings for which language is used for message translations, how numbers are displayed, what order strings get sorted into, etc.

with a few versions of R, on several operating systems, and particularly, *on someone else's machine.*

Continuous integration (CI) services make checking your package on lots of platforms much less effort. The idea is that you hook the CI service to github or Bitbucket (or whatever project hosting platform you are using), then whenever you push changes to the platform, the CI service will automatically build and check your package.

There are dozens of CI services, but not many have strong support for R. Currently the most popular services in the R community are Travis CI (https://travis-ci.org), AppVeyor (http://www.appveyor.com), and SemaphoreCI (https://semaphoreci.com). Travis CI has perhaps the most complete feature set for R packages, but it only works with packages hosted in github (no Bitbucket support). AppVeyor is mostly used because it checks packages under Windows Server (2012 at the moment), whereas Travis CI, SemaphoreCI and most other services build under Linux.[11] SemaphoreCI provides an alternative to Travis that also supports Bitbucket, though there is a tiny bit more effort in getting it set up.

> As with github and Bitbucket, you should register an account with one or more of these CI services if you plan to develop R packages. All three are free to use for public repositories.

To use a CI, visit the website and sign in to the service, follow the onscreen instructions to link up your github or Bitbucket account, and hook up the package. It's only a couple of minutes of pointing and clicking.

For Travis CI and AppVeyor, the building and checking instructions are contained in a YAML[12] configuration file. You can add these files using *devtools*.

```
use_travis("hypotenuser")
use_appveyor("hypotenuser")
```

SemaphoreCI doesn't use configuration files; rather you have to set things up using the web interface. This interface lets you write shell commands to handle checking process. Fortunately, Gabor Csardi's R-builder project (https://github.com/metacran/r-builder) has all the details covered, so you just need a few lines of boilerplate code. In the Setup section, you need two lines (always the same).

```
curl -OL https://raw.githubusercontent.com/gaborcsardi/r-builder/master/pkg-build.sh
chmod 755 pkg-build.sh
```

---

[11] As of the time of writing, there don't seem to be any free CI services that build against OS X, BSD, or Solaris.

[12] A markup language that's a bit like a human-readable version of XML. It's becoming very popular for configuration files.

Then set up two threads; the first one has these four lines of script.

```
export RVERSION=release
./pkg-build.sh bootstrap
./pkg-build.sh install_deps
./pkg-build.sh run_tests
```

The second thread script is the same, except building against the development version of R.

```
export RVERSION=devel
./pkg-build.sh bootstrap
./pkg-build.sh install_deps
./pkg-build.sh run_tests
```

Further setup details are available on the R-Builder github page.

5.5   Testing Packages on CRAN

Once your packages are mature enough to be useful to other people, you'll likely want to release them to CRAN. Mostly this doesn't involve changing anything in your package, but there are a few things to bear in mind.

- Long-running tests are frowned upon. Individual tests should take less than five seconds to run, and the total running time for all tests should be about a minute or less. The time that it takes to run your tests on CRAN can be highly variable, depending upon the load of the machines, so don't try and test the time taken to run a command.

- Your package gets tested under five different operating systems (Windows, Mac OS X, Linux, BSD, Solaris). For testing things that are operating system (OS) dependent (file paths, etc.), you can use the functionality in *assertive.reflection*. Numerical precision varies across OSes, so don't be overly ambitious with the `tol` argument to `expect_equal`.

- CRAN machines are heavily locked down. If you need to write files, you can only write to `tempdir()`. You may not even be able to see the contents of many directories.

- Your package will be tested on both the release and development version of R. You should check that your code works on both versions before submitting. See the previous section on continuous integration!

- Tests are performed with the `LANG` environment variable set to `EN` (so error messages and warnings appear in English, if you have translatable messages).

Similarly, the locale uses `LC_COLLATE = "C"`, which standardizes the sort order of text.

If you have a test that you just can't make suitable for CRAN systems, you can make it skippable on CRAN using `skip_on_cran`.

```
testthat(
  "A silly, very long running test",
  {
    skip_on_cran() # Skipping due to length of run time
    expect_true({Sys.sleep(1e6); TRUE})
  }
)
```

Clearly you shouldn't make a big habit of using this – it's far better to write tests that will work everywhere. Occasionally, the effort just isn't worth it, and it's easier to focus your development talents elsewhere. If you do use `skip_on_cran`, it's good practise to include a comment in your code as to why you are skipping the test.

There are other variant functions for avoiding tests. `skip_on_travis` and `skip_on_appveyor` will skip tests on those continuous integration platforms, and `skip_if_not_installed` will skip a test if a particular package is not installed. This is useful for other people running your tests that depend upon a *Suggested* package.

5.5.1 Testing Packages with r-hub

Part of the problem with trying to submit to packages to CRAN is that R sometimes behaves slightly differently on different operating systems. This is particularly true of anything involving locales or file systems or compiled code. CRAN tests packages on Windows, Debian and Fedora Linux, macOS, and Solaris. That means that ideally you should check that your package works under these 5 different operating systems. Actually, it's worse than that because as well as testing that your package works with the current release of R, you also need to test that the development, patched, and previous releases work too. Then there are additional complications like making sure that Solaris works with both x86 and Sparc processors. Or that C code compiles with both gcc and clang. Needless to say, nobody ever checked all the possible combinations. This has caused much frustration and friction in the past, when packages developers have had their package rejected because of a bug on Solaris that they couldn't reproduce themself.

r-hub solves this by letting you upload a package, and testing it on all operating system and release combinations, then emailing you the results. It

is free (paid for by the R Consortium[13]), and easy to use. Simply build a
source version of your package with the `build` function, then visit

    https://builder.r-hub.io

and upload your package. r-hub builds and checks your package on all the
OS/release conbinations that CRAN does, then emails you the results. This
lets you find problems in your packages without having to involve one of the
CRAN maintainers.

5.6   Calculating Test Coverage Using *coveralls.io*

As your package grows, it's easy to "just quickly add a feature," and not get
around to testing it. This is perfectly normal development behaviour,[14] but if
you do it too much, you're back to the situation of an untested package and
not knowing whether your code runs correctly or not.

In order to find out which functions need the most testing, and to get a
rough idea of how much work you need to do, you can check the level of test
coverage; Jim Hester's *covr* [15] package is the current best solution for this.

```
library(covr)
coverage <- package_coverage("hypotenuser")
percent_coverage(coverage)

# [1] 100
```

In this case, since the package is so simple, we get 100% test coverage. For
real-world situations, 100% is usually too much effort, and your time could be
better spent developing rather than testing. A good rule of thumb is that you
should have at least one test for every exported function, and the number of
tests should be proportional to how complicated the function is squared.

covr also has tools to work with online code coverage services codecov.io
and coveralls.io. The setup process is easiest with Travis CI and AppVeyor;
register for an account with the coverage service, follow the instructions in
the website to connect the package to the continuous integration service, then
edit the YAML configuration file to tell the CI service to use the coverage
service. Again, this can be done with devtools.

---

[13]Disclosure: My employer is an R Consortium member, and I have a personal project
funded by R Consortium.

[14]Unless you have way more self-discipline than I do.

```
use_coverage("hypotenuser", type = "codecov") # or "coveralls"
```

5.7 Summary

In this chapter we learned

- How to create a package with tests.

- How to use git, online code hosting, and continuous integration services.

- How to make your tests CRAN compliant.

- How to check the test coverage of a package.

6

Advanced Development-Time Testing

Congratulations! You already know everything you need to know about development-time testing... for most scenarios. This chapter contains strategies for dealing with some tricky situations that you may occasionally encounter.

6.1   Code with Side Effects

The majority of R code follows the functional programming style: you have an input, call a function, and get an output. In some cases, however, you also get *side effects* when you run your code. Typical side effects include drawing a plot, creating or modifying a file, loading a package, changing the working directory, or changing global options.

When you run tests, you need to be careful that changing the state of R doesn't affect the outcome of your other tests – it can also cause difficult-to-debug test problems, and the worst case is that your tests pass when they shouldn't. To avoid this issue, it is best practise to undo any side effects after your test finishes.

For example, suppose we want to run a test where the `stringsAsFactors` global option is set to `FALSE`. We could change the option at the start of the test, and change it back at the end, as follows:

```
test_that(
  "data.frames created with stringsAsFactors = FALSE
  have character columns",
  {
    old_ops <- options(stringsAsFactors = FALSE)
    data <- data.frame(x = LETTERS)
    expect_identical(data$x, LETTERS)
    options(old_ops)
  }
)
```

The problem with this is that, if the test throws an error part way through, the last line won't get called, and the global option will be changed for the

other tests. We can rewrite the test using `on.exit` to guarantee that the
option will be restored when the test exits. (Code inside a call to `on.exit`
is called when the current function exits, regardless of whether an error was
thrown.)

```
test_that(
  "data.frames created with stringsAsFactors = FALSE
  have character columns",
  {
    old_ops <- options(stringsAsFactors = FALSE)
    on.exit(options(old_ops))
    data <- data.frame(x = LETTERS)
    expect_identical(data$x, LETTERS)
  }
)
```

This is perfectly fine, but we can avoid writing this sort of boilerplate code
by using the *withr* [16] package, which provides a selection of functions for
temporarily introducing side effects. This next test is identical to the last one
(though I find it easier to read).

```
library(withr)
test_that(
  "data.frames created with stringsAsFactors = FALSE
  have character columns",
  {
    with_options(
      c(stringsAsFactors = FALSE),
      {
        data <- data.frame(x = LETTERS)
        expect_identical(data$x, LETTERS)
      }
    )
  }
)
```

There are a few types of side effects that aren't easily dealt with using
withr. Testing functions that write to files and testing plots are considered
later in the chapter.

6.1.1 Exercise: Writing a Test That Handles Side Effects

Temporarily set the `digits` global option to 10, and check that printing `pi`
causes the value `3.141592654` to be output to the console.

6.2   Testing Complex Objects

The expectations in *testthat* are mainly geared towards testing simple things like "are these numbers equal?" or "is this value TRUE?" Usually this is exactly what you need, but occasionally you might want to test a more complex object, like the return value from a modelling function. For example, suppose you run a linear regression with lm. How do you test the result?

There are two possible answers to this. If you are just running a regression and want to know that your answer is sensible, then you don't need to bother with unit tests – a simple assertion or two to make sure that the coefficients aren't nonsense is usually enough.

Let's take a look at a linear regression model of the *cars* dataset. This is a classic dataset from 1930 that ships with R, containing the stopping distances of cars at various speeds.

To make sure that the model is sensible, we ought to check that the model predicts the stopping distance to increase with the speed of the car.

```
model <- lm(dist ~ speed, cars)
assert_all_are_positive(coefficients(model)["speed"])
```

On the other hand, if you are creating a package that uses the complex object, you need to test its structure in detail. The structure of the lm object that is returned is rather complicated – take a look at the multi-page output from str(model) if you want to see this.

Considering the first case where you are a developer wanting to know your object is correctly generated, you could use a single call to expect_equal or expect_identical. This is fine if the test passes, but if your model object is incorrect in some way, working out what went wrong can be tricky. A better approach is to have lots of expectations that test individual elements of the object. My personal preference is to limit the number of expectations per test to two or three – otherwise I get confused. (Your limit may be higher.) For testing an object as complicated as lm, we then need multiple tests. In the following examples, note that we can define the (actual) model results outside the test code, to allow it to be reusable across the tests.

The first tests should make sure that you have an object of the correct class, and – since in R that doesn't guarantee that the object is correctly formed – that the result has the correct structure.

```
test_that(
  "the model has class 'lm'",
  {
    expect_is(model, "lm")
  }
)
```

```
test_that(
  "the elements of the model are correct",
  {
    expected <- c(
      "coefficients", "residuals", "effects", "rank",
      "fitted.values", "assign", "qr", "df.residual",
      "xlevels", "call", "terms", "model"
    )
    actual <- names(model)
    expect_equal(actual, expected)
  }
)
```

Then you can test individual elements of the result. The coefficients are perhaps the most important, so start with those.

```
test_that(
  "the coefficients are correct",
  {
    expected <- c("(Intercept)" = -17.58, speed = 3.93)
    actual <- coefficients(model)
    expect_equal(actual, expected, tol = 0.01, scale = 1)
  }
)
```

You should also perform some sanity tests on the fitted values. For example, the fitted values plus the residuals should be equal to the dependent variable in the dataset. Note the use of `expect_equivalent` to ignore the names that `fitted.values` creates.

```
test_that(
  "fitted values + residuals equal the dependent variable",
  {
    expected <- cars$dist
    actual <- fitted(model) + residuals(model)
    expect_equivalent(actual, expected, tol = 0.01, scale = 1)
  }
)
```

There are loads more possible tests, but hopefully you get the idea.

One other possible strategy is to save a variable with a known, correct structure to an RDS file, then use `expect_equal_to_reference` to test against. I don't really like this approach, since it will only tell you if the whole thing is correct or not, without telling you where things have gone wrong. If you do want to try this, the code structure looks something like the following.

First, you save your known-to-be-correct reference object. In a package context, the code for creating this should be in the **tests/testthat** directory, but named so that `test_check` won't find it. That is, the file name shouldn't start with "test."

```
saveRDS(correct_model, "tests/testthat/correct_model.rds")
```

Then, the test looks much like a standard `expect_equal` test, except that the second argument is a path to the RDS file.

```
test_that(
  "The new model equals the reference",
  {
    expect_equal_to_reference(
      new_model,
      "cars_model.rds"
    )
  }
)
```

6.2.1   Exercise: Testing a Complex Object

Perform a *Student's t-test* on the built-in *sleep* dataset using

```
test_results <- t.test(
  extra ~ group,
  data = sleep,
  paired = TRUE
)
```

1. Check that the extra hours of sleep are significantly different between the two groups of patients, with 95% confidence.
2. Write tests on the structure of the test results to ensure that the object has a suitable form.[1]

6.3   Testing Database Connections

The big problem with tests that involve databases is that the database typically lies on a different machine to your test machine. Since network connections exist outside of R, it is difficult to reliably run the test. To solve this, it is typically best to cheat a little and substitute a local data source instead.

But how do you get your code that calls a faraway database to use your alternate dataset? The trick is to overwrite the existing behaviour of your

---

[1]Yeah, this is ambiguous. Have a think about what might be important to test.

code using a *mock*. Mocking just means replacing existing functions with your own versions that cheat a bit to give you the answer that you want.[2]

Before we get to that, let's see how to structure your database-calling code. The following example uses the *DBI* [17] package to connect to a theoretical remote database. *DBI* provides a unified syntax for accessing several database management systems (DBMSs) – currently SQLite, MySQL/MariaDB, PostgreSQL, and Oracle are supported, as well as a wrapper to the Java Database Connectivity (JDBC) application programming interface (API). (Connections to SQL Server use a different system, based upon the *RODBC* package, but the principles for testing are the same.)

First we need a connection to the database. I prefer to keep this in a separate function, so you only need to change your code in one place when your IT department moves the database server.[3] Note that I'm using the *RPostgreSQL* [3] package here. *RPostgres* [30] is a newer rewrite of this package, but still seems to be a work in progress right now. The *dplyr* contains wrapper functions to connect to databases; they all have names beginning src, for example, src_postgres.

In the following example, notice a trick used to avoid storing your database password in a public place. (You **really** don't want to put your password in a public repository on github.) If you store the password in an environment variable on your machine, then it isn't available to the general public, and R still has access to it via the Sys.getenv function.

```
connect_to_db <- function()
{
  DBI::dbConnect(
    RPostgreSQL::PostgreSQL(),
    dbname   = "mydb",
    host     = "myserver",
    user     = "me",
    password = Sys.getenv("MyDatabasePassword")
  )
}
```

DBI contains a function named dbGetQuery to retrieve records from a table. To make sure that the connection is always correctly severed regardless of whether or not an error was thrown, I usually wrap that function with connection and disconnection code like so.[4]

---

[2]In object-oriented programming lanuages, the definition is a bit fancier, but this is fine for R.

[3]The function connects to a pretend PostgreSQL database, but you are welcome to imagine it is one of the other database types if it makes you feel better.

[4]Most of the time I like to connect to the database fresh every time I run a query, on the grounds that for big enough data the connection time is negligible. If you are running lots of smaller queries, this strategy will ruin your performance and you should reuse the connection.

```
get_data_from_db <- function(query)
{
  conn <- connect_to_remote_db()
  on.exit(DBI::dbDisconnect(conn))
  DBI::dbGetQuery(conn, query)
}
```

Next, you can have lots of functions for achieving specific tasks. Here's an example that retrieves some pretend revenue data from our database.

```
get_revenue_data_from_db <- function(
  start_date = end_date - 90, end_date = Sys.Date())
{
  query <- sprintf(
    "SELECT date, revenue, region
      FROM sales
      WHERE date >= '%s' AND date <= '%s'",
    start_date,
    end_date
  )
  get_sales_data_from_db(query)
}
```

Now we have a variety of options for things to mock when we create our tests. There is a tradeoff between how much code to preserve and how fast and portable we can make our code.

6.3.1   Option 1: Mock the Connection

By mocking the connection, we replace the smallest amount of code possible. This means that we get to test most of our "real" code. The downside is that, at least in this case with a PostgreSQL database,[5] you can't make the test fully portable since the database can't be included inside a package.

Next downside, there's also a bit of setup. You'll need to create another database with the same structure as the original, plus some sample data.[6]

Our *testthat* test looks the same as usual, except that the contents of the test are wrapped in a call to with_mock. This is a rather clever function that uses named arguments to override existing functions, and unnamed arguments to contain the test expectations. In the following example, we override connect_to_db (using a named argument) to connect to our local database instead of the network copy. The actual test is contained in the next argument to with_mock.

---

[5]Same with MySQL, SQL Server, Oracle, pretty much anything that isn't SQLite.

[6]Any half-decent database admin tool should let you do this without too much hassle.

```
test_that(
  "get_revenue_data_from_db returns some results",
  with_mock(
    connect_to_db = function()
    {
      DBI::dbConnect(
        RPostgreSQL::PostgreSQL(),
        dbname   = "mydb",
        host     = "localhost", # points to local db
        user     = "me",
        password = Sys.getenv("MyDatabasePassword")
      )
    },
    {
      # Some expectations to test, e.g.,
      expected <- some_results_data.frame
      actual <- get_revenue_data_from_db()
      expect_equal(actual, expected)
    }
  )
)
```

As I mentioned, creating another local database is all very well for avoiding the problem of absent network connections, but you still don't have a portable test. One variant of the above technique is to swap your network database for an SQLite or MonetDB database. These are only a single file, so you can easily include them inside a package. The two downsides to this are that your database admin tools probably won't easily let you generate an SQLite database from whatever you had already, and that SQL differs slightly from database to database, so your code might not be valid with SQLite/MonetDB. Simple "SELECT" queries are fine; anything else and you take your chances.

6.3.2   Option 2: Mock the Connection Wrapper

Remember the get_data_from_db that seemed a tiny bit frivolous? For testing, this actually gives us a great deal more flexibility. By mocking this function, you can avoid connecting to a database completely. R has some perfectly good ways of distributing data, most obviously by including data frames inside packages. This means that you can have a portable test that doesn't depend upon anything outside of R.

How do we set this up? First, you need to create a data frame for each of the tables that you want to run a query against, and use the save command to save them to file. If this code is going into a package, the variables need to be saved in the data directory.

```
sales <- data.frame(
  date = some_dates,
  revenue = some_revenues,
  region = some_regions
)
save("sales", "data/sales.RData")
```

Now the mock version of **get_data_from_db** needs to load the **sales** dataset, then call **sqldf** from the package of the same name to run the query.[7] The test containing this mock has a similar structure to the previous case.

```
test_that(
  "get_revenue_data_from_db returns some results",
  with_mock(
    get_data_from_db = function(query)
    {
      data(
        "sales",
        package = "mypkg",
        envir = parent.frame()
      )
      # or load("sales.RData") if this isn't in a pkg
      sqldf::sqldf(query)
    },
    {
      # Some expectations to test, e.g.,
      expected <- some_results_data.frame
      actual <- get_revenue_data_from_db()
      expect_equal(actual, expected)
    }
  )
)
```

This time we've made a slightly different set of tradeoffs. The benefits are that our test is now portable (since we can include the dataset inside the package), the test is faster to run (since loading RData files should be faster than retrieving data from a database), and we still get to test the SQL code that we wrote. The disadvantage is that – due to the many variants of SQL – how **sqldf** interprets the code may not be exactly the same as how your database would have interpreted it. Again, SELECT queries should be fine, but you have to be very careful with very complicated queries.

---

[7]**sqldf** is a rather nifty function that converts raw SQL queries into the equivalent R commands. Its primary use is for SQL users who have somehow stumbled into using R, but it has a few genuine use cases for R programmers too.

6.3.3   Option 3: Mock the Specific Query Functions

If retrieving the data is less of the focus of the test but simply a step needed within the test, you can mock the specific query function. For example, if you want to test a regression model of some data from the database, you can mock get_revenue_data_from_db with some locally created made-up data. In this case, all the mocked version of the function needs to do is return a data frame filtered by date.[8] Again, the structure of the test is the same: we override a function in the first argument of with_mock, and have some standard expectation code in the second argument.

```
test_that(
  "modelling revenue data gives sensible results",
  {
    with_mock(
      get_revenue_data_from_db = function(
        start_date = end_date - 90, end_date = Sys.Date()
      )
      {
        sales <- data.frame(
          date = some_dates,
          revenue = some_revenues,
          region = some_regions
        )
        sales %>%
          dplyr::filter_(
            ~ date >= start_date,
            ~ date <= end_date
          )
      },
      {
        revenue_data <- get_revenue_data_from_db()
        model <- lm(
          revenue ~ date + region,
          revenue_data
        )
        actual <- model$coefficients["date"]
        expect_greater_than(actual, 0)
      }
    )
  }
)
```

What are the tradeoffs this time? Well, we've gone further away from our "real" code in that we've replaced all the database-related functionality.

---

[8]I've used the fancy *dplyr* way of subsetting data frames because the syntax is wonderful once you are used to it. You can stick to standard-issue square brackets if you are more comfortable.

So this isn't suitable for testing the database itself. On the other hand, the test is portable and fast (since the data is local) and this is the best way to test functionality downstream from the database. That is, if you want to test functionality that would normally have used data from a database, but you don't want to worry about the database part, then use this option.

6.3.4 Summarizing the Pros and Cons of Each Database Method

Mocking the connection:

\+ You can test connecting to a real DB.

\- Having external dependencies reduces portability.

\- Connecting to a database means slower running tests.

Mocking the connection wrapper:

\+ Loading RData files is faster than connecting to a DB.

\+ No external dependencies makes the tests portable.

\+ You get to test your SQL query code.

\- You can't test the connection to an actual DB.

Mocking the specific query functions:

\+ You can focus on the downstream analysis.

\+ No DBs or SQL are involved.

\+ No external dependencies makes the tests portable.

\- You have to write more test code, since you rewrite each SQL query in R.

\- You can't test the connection to an actual DB.

\- You can't test your SQL queries either.

6.4 Testing Rcpp Code

R is mostly optimised for fast writing of code and fast reasoning about code.[9] The tradeoff is that sometimes the execution speed of your code isn't as fast

---

[9]At least in theory.

as you'd like it to be. When you need fast execution, it can be beneficial to switch to a compiled language: R supports calling C, C++, and FORTRAN code. Of these, calling C++ code is by far the easiest, thanks to the *Rcpp* package.

C++ has a bit of a reputation for being tricky to master. The good news is that, for data analysis purposes, you can get a lot of benefits without spending years becoming a master. The two big use cases are

- Avoiding struggling to vectorize code: loops in R are slow, but fast in C++. If vectorized R code is eluding you, a couple of `for` loops in C++ can often work magic.

- Accessing existing code: there are lots of high-quality numerical libraries in C++.

Needless to say, this is a huge topic, and only a very brief snippet is covered here. The definitive guide is Dirk Eddelbuettel's *Seamless R and C++ Integration with Rcpp* [10]. There is a lot of sample code available from http://gallery.rcpp.org, where you can learn by example, and Hadley Wickham's *Advanced R* [24] has a chapter introducing the contents.

6.4.1 Getting Set Up to Use C++

Before you begin, you need to make sure that you have the *gcc* and *make* tools installed on your machine, in a place where R can find them. On a Unix-based operating system, these are installed as standard. Under Windows, you should install Rtools first; see the Prerequisites section in Chapter 5 for details. You can double check that these tools are available using `assert_r_can_compile_code`.

```
library(assertive.reflection)
assert_r_can_compile_code()
```

To add the infrastructure necessary for using *Rcpp* to a package, *devtools* has a function `use_rcpp`.[10] This creates a `src` directory and updates the `DESCRIPTION` file to import and link to *Rcpp*.

```
library(devtools)
use_rcpp("path/to/your/package")
```

`use_rcpp` will also give you a message that you need to have some roxygen tags in your R code. You can either manually add those lines, or use `writeLines` like so:

---

[10]`Rcpp.package.skeleton` in the *Rcpp* package is loosely equivalent to `create` + `use_rcpp`.

```
writeLines(
  "#' @useDynLib yourpkg
#' @importFrom Rcpp sourceCpp
NULL",
  "path/to/your/package/R/rcpp-setup.R"
)
```

Each of your C++ functions will have an R-level wrapper function. This is what you or your users will call. That means that it is perfectly possible to get away with testing everything at the R level, and these tests can be performed in exactly the same way as any other tests.[11] For cases where the C++ code is complex, you can also test the C++ code directly using the *Catch* unit testing framework. *testthat* has a function named use_catch to add the required infrastructure to your package. This creates three files:

src/test-runner.cpp This makes your package export a C routine, run_testthat_tests, which lets *testthat* run Rcpp tests. Don't touch this file; just leave it to work its magic.

tests/testthat/test-cpp.R This just calls expect_cpp_tests_pass which was briefly mentioned in Chapter 3. Again, don't touch this.

src/test-example.cpp This creates an example file that demonstrates how to use Catch tests. Read it, understand it, delete it.

```
library(testthat)
use_catch("path/to/your/package")
```

use_catch will also ask you to edit the package's DESCRIPTION file to link to the *testthat* package. You can manually do this, or there is an internal *devtools* function that will do it for you,[12] named add_desc_package.

```
devtools:::add_desc_package(
  "path/to/your/package", "LinkingTo", "testthat"
)
```

You can keep Catch unit tests in the same file that you define the functions you want to test. These C++ files (with file extension .cpp) have a reasonably standard structure that is worth exploring.

The C++ files should start by importing any libraries that they will use using include statements. You need #include <Rcpp.h> to use *Rcpp*, and if you are including any tests, you also need #include <testthat.h>. There are lots of C++ libraries that you may also wish to include, for example, to access many fundamental mathematical functions, you'll also need to #include <math.h>.

---

[11]Since this is easier, it should be your first choice strategy.

[12]If you want to do this programmatically but don't want to rely on an internal function, then use read.dcf and write.dcf in the *base* package, or specify this link when you call create.

It is also useful to import the *Rcpp* namespace with a `using` command, to save having to prefix variable types with `Rcpp::`.

```
#include <Rcpp.h>
#include <testthat.h>
using namespace Rcpp;
```

Next we provide function definitions. A lot of C++ code looks fairly similar to R code;[13] but there are a few important syntactical differences. Here's a function that calculates the geometric mean.

```
// [[Rcpp::export]]
double geomean(NumericVector x) {
  return exp(mean(log(x)));
}
```

These elements of the code block are worth considering:

`// [[Rcpp::export]]` This line is a C++ attribute that declares that we want to export our function. That is, *Rcpp* will make an R-level wrapper function for users to call.

`double` C++ functions must declare the type of value that they return. `double` means a single double-precision floating-point number, equivalent to a length one `numeric` vector in R.

`NumericVector` C++ functions must also declare the type of each input argument. A `NumericVector` is an *Rcpp* type that corresponds to a `numeric` vector in R.

`return` The return value from a function must be explictly declared using `return`. (This isn't optional, as in R.)

`exp(mean(log(x)))` *Rcpp* contains "sugar" functions that behave like their R counterparts. These work on `NumericVectors`.

`;` All code statements in C++ must finish with a semicolon.

Catch unit tests look very similar to *testthat* tests, except that you only have a choice of `expect_true`, `expect_false`, `expect_error`, and `expect_error_as`.[14]

In the following code snippet, the *Rcpp* sugar function, `all` returns a `LogicalVector`. Catch doesn't understand these, so we have to wrap it in a call to the *Rcpp* function `is_true`.[15]

---

[13]Give or take your code style, of course.

[14]C++ has a hierarchy of exception classes. `expect_error_as` lets you test the type of error, whereas `expect_error` tests for any error.

[15]Not to be confused with `assertive.base::is_true` or `testthat::is_true`. It's a common function name!

```
context("test geomean") {
  test_that("the geomeans are correct") {
    NumericVector x = NumericVector::create(1.0, 2.0, 3.0, 4.0, 5.0);
    double expected = 2.6051710846973517;
    NumericVector actual = geomean(x);
    expect_true(is_true(all(actual == expected)));
  }
}
```

6.4.2   Case Study: Extending the *hypotenuser* Package to Include C++ Code

Let's extend the *hypotenuser* package that we created in Chapter 5 to include a C++ version of the hypotenuse and some tests for it.

> If you want to follow along with these step, rather than typing everything yourself, visit the online version of the package:
>   https://bitbucket.org/richierocks/hypotenuser
> By viewing the Commits page, you can see how the package changes at each step.

First we need to add the infrastructure to the package.

```
# Adding Rcpp to LinkingTo and Imports
# * Creating 'src/'.
# * Ignoring generated binary files.
# Next, include the following roxygen tags somewhere in your package:
#
# #' @useDynLib hypotenuser
# #' @importFrom Rcpp sourceCpp
# NULL
#
# Then run document()
# > Added C++ unit testing infrastructure.
# > Please ensure you have 'LinkingTo: testthat' in your DESCRIPTION.
# > Please ensure you have 'useDynLib(hypotenuser)' in your NAMESPACE.


use_rcpp("hypotenuser")
use_catch("hypotenuser")
```

We also need to create a file named "R/rcpp-setup.R" containing the lines

```
#' @useDynLib hypotenuser
#' @importFrom Rcpp sourceCpp
NULL
```

Additionally, we need to add *testthat* to the packages in the `LinkingTo` field of the `DESCRIPTION` file.

Next we need to add a C++ file to the `src` directory that defines our hypotenuse function.

```cpp
#include <Rcpp.h>
#include <testthat.h>
#include <math.h>
using namespace Rcpp;

// [[Rcpp::export]]
NumericVector hypotenuseCpp(NumericVector x, NumericVector y) {
  return sqrt(x * x + y * y);
}
```

Although the `[[Rcpp::export]]` attribute will cause an R-level function to be generated, we still need to write some roxygen documentation for it, including an `@export` tag.

```r
#' Calculate hypotenuses
#'
#' Calculates the hypotenuse, using C++.
#' @param x A numeric vector.
#' @param y Another numeric vector.
#' @return The hypotenuse.
#' @name hypotenuseCpp
#' @export
NULL
```

In this case, the most sensible option is to do the testing of `hypotenuseCpp` at the R level. In this way, we can write tests in the same way as for any other function. That is, in the package's `tests/testthat` directory, we add something like the following:

```r
test_that(
  "hypotenuseCpp works at the R level",
  {
    x <- c(3, 5, 8)
    y <- c(4, 12, 15)
    expected <- c(5, 13, 17)
    actual <- hypotenuseCpp(x, y)
```

```
    expect_equal(actual, expected)
  }
)
```

At the C++ level, we can add a Catch test to the existing .cpp file.

```
context("test hypotenuseCpp") {
  test_that("hypotenuseCpp works at the C++ level") {
    NumericVector x = NumericVector::create(3.0, 5.0, 8.0);
    NumericVector y = NumericVector::create(4.0, 12.0, 15.0);
    NumericVector expected = NumericVector::create(5.0, 13.0, 17.0);
    NumericVector actual = hypotenuseCpp(x, y);
    expect_true(is_true(all(actual == expected)));
  }
}
```

Finally, we roxygenize, build and check the package as usual.

```
roxygenize("hypotenuser")
build("hypotenuser")
check("hypotenuser")
```

6.4.3 Exercise: Testing an *Rcpp* Function

Write a C++ file containing a function that calculates the square of the input. Write a Catch unit test for this function. Integrate this into a package. Your function should accept a NumericVector input and return a NumericVector too. Note that there is no ^ operator in C++, so you can calculate the square using x * x, or pow(x, 2.0). You don't need to use a loop.

6.5 Testing Write Functions

Functions that write data to a file can be a little messy to deal with, since you have to start testing files rather than R variables, you need to make sure that the requisite files exist, and you need to clean up after yourself.

If you are testing a function that writes text output, and that works with *connections*, the best option is to write to stdout (that is, to the console), capture it with capture.output, then test the resulting character vector.

```
test_that(
  "write.csv produces the correct output",
  {
    data <- data.frame(x = 1:3, y = letters[1:3])
    # output has 3 columns, including row names
    # strings are quoted with double quotes
    expected <- c(
      '"","x","y"',
      paste(shQuote(1:3), 1:3, shQuote(letters[1:3]), sep = ",")
    )
    actual <- capture.output(write.csv(data, file = stdout()))
    expect_equal(actual, expected)
  }
)

# Error: Test failed: 'write.csv produces the correct output'
# * 'actual' not equal to 'expected'.
# 3/4 mismatches
# x[2]: "\"1\",1,\"a\""
# y[2]: "'1',1,'a'"
#
# x[3]: "\"2\",2,\"b\""
# y[3]: "'2',2,'b'"
#
# x[4]: "\"3\",3,\"c\""
# y[4]: "'3',3,'c'"
```

Note that using `expect_output` isn't possible here, since you can only match against a single regular expression, so the multiple lines of output aren't supported.[16]

For functions that write binary files, you can't write to stdout, so typically you need to test the result against a correctly formed file. To compare files, the best way to create hashes of the files, and compare those.[17] We'll use the `digest` function from the package of the same name [31] for this. The next example tests `saveRDS`, which is used for saving R variables to a binary file.

```
library(digest)
test_that(
  "saveRDS produces the correct output",
  {
    data <- data.frame(x = 1:3, y = letters[1:3])
```

---

[16]Without a lot of messing about.

[17]A hash function takes a complex object and reduces it to a string. The probability of two different objects generating the same string is very low, so you can assume that equal hashes mean identical objects.

```
  # create expected output file
  expected_file <- file.path(tempdir(), "expected.rds")
  con <- gzfile(expected_file, open = "wb")
  .Internal(
    serializeToConn(data, con, FALSE, NULL, NULL)
  )
  close(con)
  # create actual output file
  actual_file <- file.path(tempdir(), "actual.rds")
  saveRDS(data, actual_file)

  # compare hashes of the files using digest
  expected <- digest(file = expected_file)
  actual <- digest(file = actual_file)
  expect_equal(actual, expected)
}
)
```

If you were looking closely at the previous block of code, you may have spotted that the test calls R's internal `serializeToConn` function. This is one of the dilemmas with trying to create a correctly formed binary file to test against – you often end up rewriting the internals of the function you want to test.[18] Sometimes it's OK; sometimes it just makes it harder to change the function, since you need to change the test in parallel.

If you have a *read* function to match your write function, then one way around this is to test both functions at the same time. Write an object to file, then read it back in, and test that you have the same object. Clearly this is a weaker test, since you aren't testing the intermediate step, and you can trivially fool the test by making both read and write the `identity` function, but if you are out of other ideas, it's a good start.

6.5.1   Exercise: Writing INI Configuration Files

INI is a text file format traditionally used for configuration files.[19] Lines consist of `name=value` pairs, split into `[section]`s.

Write a function that accepts a list of named atomic vectors and write it in INI format. For example, the input

```
ini <- list(
  section1 = c(element1 = 1, element2 = "a"),
  section2 = c(element3 = TRUE)
)
write_ini(ini, "some_file.ini")
```

---

[18]The presence of an internal function also means that this test wouldn't be allowed on CRAN; you'd have to add a call to `skip_on_cran`.

[19]These days YAML is preferred instead.

should write the following output to file.

```
[section1]
element1=1
element2=a

[section2]
element3=TRUE
```

Write a test for this function to make sure it works.

Don't worry about having trailing blank lines when you write to file or run your test.

6.6   Testing Graphics

Testing that your plots look OK is a little different from other test types, because you really need to physically look at them in order to make sure that they are drawn correctly.[20] For this reason, testing graphics isn't completely automatable, but there are steps that you can take in order to make it reproducible and relatively painless.

The *knitr* package allows you to create documents that mix text, images, code, and the output from that code together. You can write them in a variety of markup languages: LaTeX, HTML, markdown, Asciidoc, reStructuredText, and textile are currently supported. It's very powerful – this book is written using knitr, for example. The best reference on the subject is Yihui Xie's *Dynamic Documents with R and knitr* [32], and there is excellent documentation available on Yihui's website, yihui.name/knitr.

For most scenarios, the markdown markup language is the best choice for writing short documents, including graphics test documents. markdown solves the problem that HTML is fiddly to write, what with all the angle brackets, and hard work for humans to read (again because of all those angle brackets). As an example, to include bullet lists in an HTML document, it's slightly fiddly.

```
<ul>
  <li>an item in the list</li>
  <li>another item</li>
</ul>
```

---

[20]A different approach is taken by Winston Chang's *vtest* [2] package, available using `devtools::install_github("wch/vtest")`. This takes hashes of plots so you can see which ones have changed between commits in your code. It's very clever but a bit fiddly, so for most purposes I find *looking at plots* to be better. *vtest* scales better to the case of having thousands of plots.

With markdown, you just start each line with a dash.

```
- an item in the list
- another item
```

To include a chunk of code in your (pre-knitted) markdown document, use three backticks, followed by "r" and a label for the chunk, in braces. (The chunk label is optional, but helps the caching system, which make things go faster.)

```
```{r, ChunkLabel}
# R code goes here
```
```

A document to test a plot should contain the text description of what you expect the plot to contain, the code for the plot, and the plot itself. Here's an example using the built-in *Orange* dataset, which details the circumference of orange trees based on their age. Lines starting # and ## represent top-level and second-level headings, respectively.

This generates a report, shown on the next page.

Tests of plots of the *Orange* dataset

Test scatter plot

- The plot should be a scatter plot

- Age should be on the x axis

- Circumference should be on the y axis

- Each tree should be given a different shape

- Ages and circumferences should all be positive values

- Circumference should increase with age for each tree

```
library(ggplot2)
ggplot(Orange, aes(age, circumference, shape = Tree)) +
  geom_point()
```

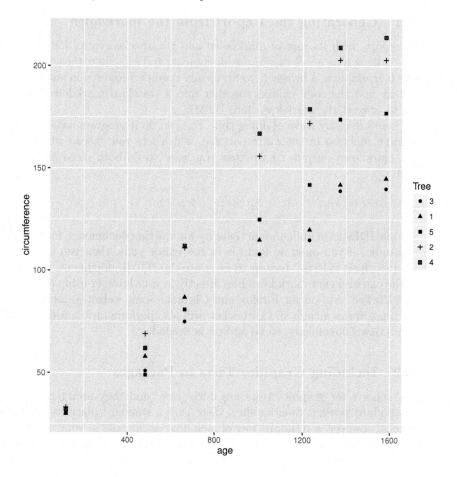

6.6.1   Generating the Report from the *markdown*

The report, with its mix of markdown and the special syntax for chunks of code, is known as an *R markdown* document, and has file extension Rmd. To make it readable in a browser, you typically need two conversion steps: to *knit* the text and the code chunks together into a standard markdown file, and then to convert the markdown into HTML.

There are many ways of doing this. You can do it programmatically using the knit function in the *knitr* package, which lets you choose whether you want markdown output, or whether you want to do both steps and output HTML.

```
knit("path/to/your/file", "outputfile.html")
```

Some IDEs have built-in functionality for knitting documents; for example, in RStudio, if you open an Rmd file in the editor pane, then you can simply click the "Knit HTML" button to generate the HTML document.

You can also view markdown files directly in your browser using the "Markdown Viewer" add-on for Firefox and Chrome. Some websites, including the code repositories github and Bitbucket, will also perform automatic rendering of markdown documents, so no add-on is needed.

6.6.2   Including Graphics Tests in Packages

Since reports for graphics tests are fairly rare, and they aren't covered by the standard *testthat* functionality, there isn't a standard place to put them. There is, however, a standard way of including documents in your package, as *vignette*s. These are usually used for describing how to use the package, or for case studies, but there's absolutely no reason why you can't include graphics tests in them as well.

To set up your package to use *knitr* vignettes, use the use_vignette function in *devtools*.

```
library(devtools)
use_vignette("path/to/your/package")
```

6.6.3   Exercise: Write a Graphics Test Report

Write a test report for a scatter plot of *weight* versus *height* for the women in the dataset of the same name.

6.7 Summary

What we learned.

- For code with side effects, use the *withr* package to undo the side effect afterwards.

- For complex objects, only test the bits that you need to.

- To test databases, or other "risky" data sources, use mocks.

- *testthat* can test C++ code, using the *catch* framework.

- Write functions can be tested by writing to the console and using `capture.output`.

- Use *knitr* to create a report for testing graphics.

7

Writing Your Own Assertions and Expectations

Chapter 2 was pretty long, right? While *assertive* is getting pretty big, there are infinitely many possible assertions. I don't have time to write an infinitely large package, so this chapter explains how to create your own custom assertions, for the things I miss. Similarly, you can create your own expectations for the *testthat* package.

7.1   The Quick, Nearly Good Enough Option

Remember that a predicate is just any function that returns a logical value. In this book, where *assertive* is concerned, we've made a bit of effort to distinguish *scalar* predicates that return a single logical value, and *vector* predicates that return a logical vector.

The `assert_engine` function in *assertive.base* takes a predicate (of either type), and creates an assertion from it. To use `assert_engine`, you pass a predicate into the first argument, and subsequent arguments are passed to that predicate. For example:

```
assert_engine(is.numeric, 1:5) # OK
assert_engine(is.numeric, letters)

# Error in eval(expr, envir, enclos): is.numeric :
#   The assertion failed.
# [1] FALSE
```

Here the first line silently ran its check and passed, and the second line threw an error. So we have a working assertion.

> Try using `assert_engine` with a predicate that returns a logical vector (rather than a single value).

The big problem with the previous example is that the error message isn't very good. You can fix this by passing the `msg` argument.

```
assert_engine(
  is.numeric,
  letters,
  msg = "letters is not numeric"
)

# Error in eval(expr, envir, enclos): is.numeric :
#    letters is not numeric
# [1] FALSE
```

The next few sections are about refining this idea to make your predicates and assertions even more user friendly.

7.2   Writing Scalar Predicates

Scalar predicates have a fairly standard layout, which you can copy to write your own. Let's take a look at the has_rows function in *assertive.properties*.

```
has_rows

# function (x, .xname = get_name_in_parent(x))
# {
#     nrowx <- nrow(x)
#     if (is.null(nrowx)) {
#         return(false("The number of rows in %s is NULL.", .xname))
#     }
#     if (nrowx == 0L) {
#         return(false("The number of rows in %s is zero.", .xname))
#     }
#     TRUE
# }
# <environment: namespace:assertive.properties>
```

The function takes two inputs, x and .xname.

x is the object that the user wants to perform the check on.

.xname represents the name of x when it is passed into the function. For example, if you called has_rows(y), then .xname would be y. The user should never have to explicitly use this argument, but it helps make the error messages a little bit clearer in the event of things going wrong.

The body of the function mostly consists of code that states "if something goes wrong, then return FALSE, with an explanation." In this function, there are two ways that x can fail to have rows – nrow(x) can be NULL, or it can be zero. If both of those bullets are dodged, then the function returns TRUE. Before we get to that, the `false` function, from *assertive.base*, requires some explanation.

`false` always returns the value FALSE, with two embellishments. Firstly, the arguments to the function are passed to `sprintf`, which forms them into a single string, which is then stored in an attribute named `cause`. `sprintf`[1] is a function inherited from C that combines text and variables into a character vector. The first argument can contain placeholders, such as %s (representing a string) in the `has_rows` function. Further arguments are substituted into those placeholders. Other common placeholders are %f for numeric values and %d for integers.

The second embellishment is to assign the FALSE value the class `scalar_with_cause`, which allows for prettier printing of the output and is necessary to make the corresponding assertion work.

There is also an `na` function, which works the same way as `false`, except that it returns NA.

7.2.1 Exercise: Writing a Custom Scalar Predicate

Write a predicate that accepts a `numeric` vector (`integer` and `logical` inputs are OK too) and returns TRUE if the vector is monotonically increasing. That is, each value is greater than or equal to the previous value.

Hint: Take a look at the `diff` function in the `base` package.

For bonus points, write some tests to make sure that this function is correct.

7.2.2 Writing Type Predicates

For testing if an object is a particular class, writing the predicate is even easier. Since there are lots of these, all the logic has been outsourced to a function named `is2` in *assertive.types*.

```
is_character

# function (x, .xname = get_name_in_parent(x))
# {
#     is2(x, "character", .xname)
# }
# <environment: namespace:assertive.types>
```

The function signature is the same: it needs x and .xname arguments, but

---

[1] Full documentation is, of course, on the ?sprintf help page, where the examples show plenty of tricks for its use.

the function body is simply a call to is2, with the class that needs to be checked.

7.3   Writing Scalar Assertions

The assertion corresponding to a scalar predicate should just be a single line of code calling the **assert_engine** function. As we saw earlier in the chapter, this calls the predicate and throws an error if that doesn't return TRUE. Consider assert_has_rows.

```
assert_has_rows

# function (x, severity = getOption("assertive.severity", "stop"))
# {
#     assert_engine(has_rows, x, .xname = get_name_in_parent(x),
#         severity = severity)
# }
# <environment: namespace:assertive.properties>
```

The assertion takes two arguments, x, the input to be checked, and severity, which controls how severe the response should be if the check fails. All the assertions in the assertive package take the default for this from the "assertive.severity" global option (which is set to "stop" when *assertive.base*) is loaded, or "stop" if the user has unset this option.

The first argument passed to **assert_engine** is the name of the predicate, in this case has_rows.

Next are the arguments to be passed to the predicate, x and .xname. Note that we need to set .xname to be the name of x in the parent environment of the assertion – we can't just use the default value in the predicate.[2] Don't worry about the technicalities, you just need to remember to include the line .xname = get_name_in_parent(x) in all your assertions.

The final argument to pass to **assert_engine** is severity.

To avoid argument bugs and assist with code clarity, I usually pass all the arguments to **assert_engine** except the first two by name.

7.3.1   Exercise: Writing a Custom Scalar Assertion

Write an assertion to correspond to the predicate testing for monotonicity that you wrote in the previous exercise.

---

[2]This is because the parent environment of the predicate is the environment of the assertion, not the environment of the calling function.

7.4 Writing Vector Predicates

As with the scalar predicates, there is a fairly standard format for vector predicates. To demonstrate how they work, take a look at is_imaginary in *assertive.numbers*.

```
is_imaginary
```

```
# function (x, tol = 100 * .Machine$double.eps,
#     .xname = get_name_in_parent(x)).
# {
#     if (!is.numeric(x)) {
#         x <- coerce_to(x, "complex", .xname)
#     }
#     call_and_name(function(x) {
#         ok <- abs(Re(x)) < tol
#         set_cause(ok, "real")
#     }, x)
# }
# <environment: namespace:assertive.numbers>
```

The meat of this function is the line `ok <- abs(Re(x)) < tol`. This returns TRUE whenever the real component of x is zero, give or take numerical rounding errors. In general, vectorized predicates will have a similar line that takes a vector input and returns a logical vector of the same length.

There are two enhancements[3] to make the function more useful. Firstly, set_cause sets the cause attribute of the return value and gives it the class "vector_with_cause." You can see how it works with a simple example.

```
set_cause(c(TRUE, FALSE, NA), "some reason")
```

```
# There were 2 failures:
#   Position Value        Cause
# 1        2        some reason
# 2        3            missing
```

There are a few things worth noting about this function. The cause of failure for cases where the input is FALSE will usually be a single string. It is possible to pass it a vector the same length as the first argument in case you need multiple reasons for failure. Cases where the input is NA are given a cause of failure of "missing," though it is possible to override this.

The second enhancement to is_imaginary is the use of call_and_name. If you look carefully at the output from the previous set_cause example, you'll notice that the value column is blank. The print method for

---

[3]Traditional British cuisine often consists of "meat and two veg," so think of these functions as the vegetables, if it helps.

`vector_with_cause` objects takes the `value` column from the names of the object.

`call_and_name` is a utility function that calls another function and names the return value with the first input. The first argument is the function to call, and further arguments are passed to that function.

```
call_and_name(log2, c(1, 2, 4, 8))
```

```
# 1 2 4 8
# 0 1 2 3
```

To summarize, we had a line of code that returned a logical vector the same length as `x`, added a `cause` attribute with `set_cause`, and named the output using `call_and_name`.

7.4.1   Exercise: Writing a Vector Predicate

DNA sequences are strings consisting of only the letters `"A"`, `"C"`, `"G"`, and `"T"`. Write a predicate for matching DNA sequences. Hint: You can do string matching with base-R's `grepl` function, or, if you prefer, using the `stringi` [12] package's `stri_detect_regex`.[4] Both these functions match regular expressions; learning to use them is beyond the scope of this book.[5] For now, know that you can match the DNA sequences using the regular expression

```
dna_regex <- "^[ACGT]+$"
```

The caret matches the start of the string, the square brackets indicate a group of characters, the plus means "match the previous thing one or more times," and the dollar matches the end of the string.

7.5   Writing Vector Assertions

Just as in the scalar case, vector assertions make a call to assert engine; the bodies of these functions are very similar. Let's take a look at the "all" assertion that corresponds to `is_imaginary` that we previously discussed.

```
assert_all_are_imaginary
```

```
# function (x, tol = 100 * .Machine$double.eps, na_ignore = FALSE,
#     severity = getOption("assertive.severity", "stop"))
```

---

[4]Or `str_detect` from the *stringr* [26] package. Whatever you like, really.
[5]But you absolutely should learn to use them at some point. www.regular-expressions.info is a good place to start.

```
# {
#      .xname <- get_name_in_parent(x)
#      msg <- gettextf("%s are not all imaginary (tol = %g).", .xname,
#          tol)
#      assert_engine(is_imaginary, x, tol = tol, .xname = .xname,
#          msg = msg, na_ignore = na_ignore, severity = severity)
# }
# <environment: namespace:assertive.numbers>
```

The first two arguments to the function, x and tol, are passed on to is_imaginary. The last argument, severity, controls how severe the consequences of failure are, as discussed in the section on scalar assertions. The na_ignore argument is new and worth a little discussion.

Some (but not all) predicates can return NA values. For example, if you call is_imaginary with an NA input, then you don't know if the value is imaginary or not, so the result should also be NA. By default, na_ignore = FALSE, missing values will cause the assertion to fail. If that isn't what you want, you can change this argument to ignore the missing values. It works like the na.rm argument to mean and other base-R numeric functions, except that the index of the failing element is preserved in the error message.

The first line of code, which creates the failure message, calls a function named gettextf. This works exactly like its more common cousin, sprintf, except that the resulting string is translatable.[6] How to translate messages is beyond the scope of this book, and if you aren't bothered about making your messages available in multiple languages, then you can stick to sprintf or paste or whatever is easiest for you.

The call to assert_engine just passes in all the arguments that we have: the four arguments that were passed into assert_all_are_imaginary, plus the msg that we just defined.

The "any" assertion is almost identical. Can you spot the two differences?

```
assert_any_are_imaginary

# function (x, tol = 100 * .Machine$double.eps, na_ignore = FALSE,
#      severity = getOption("assertive.severity", "stop"))
# {
#      .xname <- get_name_in_parent(x)
#      msg <- gettextf("%s are never imaginary (tol = %g).", .xname,
#          tol)
#      assert_engine(is_imaginary, x, tol = tol, .xname = .xname,
#          msg = msg, what = "any", na_ignore = na_ignore,
#          severity = severity)
# }
# <environment: namespace:assertive.numbers>
```

---

[6] *assertive.base* currently has translations into Dutch, French, German, Greek, Hungarian, Korean, Russian, Swedish, Turkish, and Ukranian. As the other packages become stable, I hope to have those translated too.

Firstly, the failure message is subtly different. We need to say *nothing* passed the check, rather than *not everything* passed the check.

There is one additional argument here: you need to pass `what = "any"`. (In the "all" case, `what = "all"` is implicitly passed.) And that's it.

7.5.1 Exercise: Writing a Vector Assertion

Write the corresponding "all" and "any" vector assertions to go with the DNA sequence matching predicate that you wrote in the last exercise.

7.6 Creating Custom Expectations

It doesn't happen often, but you may occasionally find that, with the existing set of expectations in *testthat*, your tests look a little clunky. In the same way that *assertive* allows you to define your own predicates and assertions, *testthat* allows you to define your own expectations. Note that, as of the current version of *testthat*, 1.0.2, this process isn't officially supported, since one of the functions you need, `make_label`, is internal. That is, you can't include tests using these expectations on CRAN, and the behaviour may change in future releases.

Suppose you want to write tests for objects that aren't `NULL`. You could write tests with lines like

```
expect_false(is.null(x))
```

This is OK, and if you just have a few tests like this, it's perfectly acceptable. If you are doing a lot of tests like this, it's worth making the effort to have more readable tests like

```
expect_not_null(x)
```

Let's take a look at how to construct `expect_not_null`. Expectations need to take at least three arguments: an `object` to test,[7] an optional `info` argument to add extra information in the failure message, and thirdly an optional `label` argument that names the object.

The body of the function is typically three lines long. First, you need to call `make_label` to get a default name for the case when `label` has not been set. Secondly, you need to call `expect`, passing the three function inputs, and a predicate. Finally, the first input object is invisibly returned, so that expectations can be piped together. The code looks like this

[7]Some of them, like `expect_equal`, need to compare two objects.

```
expect_not_null <- function(object, info = NULL, label = NULL)
{
  lab <- testthat:::make_label(object, label)
  expect(!is.null(object), sprintf("%s is null.", lab),
      info = info)
  invisible(object)
}
```

Let's give it a try. This test should pass

```
test_that(
  "NA is not NULL",
  {
    expect_not_null(NA)
  }
)
```

...and this silly test should fail.[8]

```
test_that(
  "NULL is not NULL",
  {
    expect_not_null(NULL)
  }
)
```

```
# Error: Test failed: 'NULL is not NULL'
# * NULL is null.
```

Mostly, you can just accept that a test fails on purpose. If you really want to be thorough and test that the test works as expected, it is possible to create a meta-test.[9]

```
test_that(
  "expect_not_null fails when there are NULLs",
  {
    expect_error(
      test_that(
        "NULL is not NULL",
        {
          expect_not_null(NULL)
        }
      )
    )
```

---

[8]If, for some reason, you want a test that fails on purpose, you can include `fail()` in the test's body.

[9]Since we're at the end, treat yourself to an XKCD web-comic. https://www.xkcd.com/917

```
  }
)
```

7.6.1   Exercise: Create a Custom Expectation

Create a custom expectation, `expect_positive`, that tests for a numeric vector containing only positive numbers.

Devise tests to make sure your expectation works.

7.7   Summary

In this chapter you learned

- How to create your own predicates and assertions.

- ...for both the scalar and vector cases.

- How to create your own custom expectations.

   Congratulations! You now know how to test R code!

A

Answers to Exercises

A.1 Preface

A.1.1 Exercise: Are You Ready?

There are lots of ways of creating a function to count the number of numeric columns. Here's a simple implementation.

```
count_numeric_cols <- function(x)
{
  col_is_numeric <- vapply(x, is.numeric, logical(1))
  sum(col_is_numeric)
}
```

This implementation doesn't check that the input is a data frame – we'll discuss how to check for this in Chapter 2.

A.2 Chapter 2

A.2.1 Exercise: Using Predicates and Assertions

The vector to check:

```
(x <- c(0, 1, Inf, -Inf, NaN, NA))
```

```
# [1]    0    1  Inf -Inf  NaN   NA
```

Check for being numeric (predicate first, then assertion):

```
is_numeric(x)
```

```
# [1] TRUE
```

```
assert_is_numeric(x) # No output, since the check passes
```

Check for being finite:

```
library(assertive.types)
is_finite(x)

# There were 4 failures:
#   Position Value        Cause
# 1        3   Inf     infinite
# 2        4  -Inf     infinite
# 3        5  <NA> not a number
# 4        6  <NA>      missing

assert_all_are_finite(x)

# Error in eval(expr, envir, enclos): is_finite : x are not all
#   finite.
# There were 4 failures:
#   Position Value        Cause
# 1        3   Inf     infinite
# 2        4  -Inf     infinite
# 3        5  <NA> not a number
# 4        6  <NA>      missing
```

Check for having no missing values:

```
library(assertive.base)
is_not_na(x)

# There were 2 failures:
#   Position Value   Cause
# 1        5    NA missing
# 2        6    NA missing

assert_all_are_not_na(x)

# Error in eval(expr, envir, enclos): is_not_na :
#   The values of x are sometimes NA.
# There were 2 failures:
#   Position Value  Cause
# 1        5    NA missing
# 2        6    NA missing
```

A.2.2 Exercise: Examining an Object

To check that the dataset is a list:

```
library(assertive.types)
assert_is_list(Harman23.cor)
```

To check that it is of length three:

```
library(assertive.properties)
assert_is_of_length(Harman23.cor, 3)
```

There are three checks here, for the class being "matrix," for row names, and for column names.

```
assert_is_matrix(Harman23.cor$cov)
assert_has_rownames(Harman23.cor$cov)
assert_has_colnames(Harman23.cor$cov)
```

To check for values between zero and one there are two options. We can use assert_all_are_in_range, or the more specific assert_all_are_proportions.

```
assert_all_are_proportions(Harman23.cor$cov)
```

To check for values being exactly zero, we also have two options. Either assert_all_are_equal_to with a tolerance of zero.

```
assert_all_are_equal_to(Harman23.cor$center, 0, tol = 0)
```

A.2.3   Exercise: Inspecting Files and Directories

Here are the contents of the root of the base package:

```
contents_of_base_pkg_dir
```

```
# [1] "/Library/Frameworks/R.framework/Resources/library/base/CITATION"
# [2] "/Library/Frameworks/R.framework/Resources/library/base/demo"
# [3] "/Library/Frameworks/R.framework/Resources/library/base/DESCRIPTION"
# [4] "/Library/Frameworks/R.framework/Resources/library/base/help"
# [5] "/Library/Frameworks/R.framework/Resources/library/base/html"
# [6] "/Library/Frameworks/R.framework/Resources/library/base/INDEX"
# [7] "/Library/Frameworks/R.framework/Resources/library/base/Meta"
# [8] "/Library/Frameworks/R.framework/Resources/library/base/R"
```

We can simply run the five file and directory checks from *assertive.files*. Note that, if you run the checks for readability/writability/executability under Windows, you'll see a warning about file.access not being reliable under Windows. For example, is_readable_file will sometimes return FALSE with a file that can be read perfectly well with read.csv. You can turn the warning off with the argument warn_about_windows = FALSE.

Directories are considered executable (at least under Windows), but text files are not.

A.2.4   Exercise: Testing Properties of Matrices

The checks for real numbers and numbers in a range come from *assertive.numbers*. In the check for a symmetric matrix, the numbers really ought to be identical (no numerical error), so we can set the tolerance to zero.

```
library(assertive.matrices)
x <- cor(longley)
assert_all_are_real(x)
assert_all_are_in_closed_range(x, -1, 1)
assert_is_symmetric_matrix(x, tol = 0)
```

A.2.5   Exercise: Explore Your R Setup

You can check for a current version of R using

```
assert_is_r_current()
```

A.2.6   Exercise: Checking Customer Data

Checking the "ID" field for positive whole numbers with no duplicates.

```
assert_is_numeric(customer_data$Id)
assert_all_are_positive(customer_data$Id)

# Error in eval(expr, envir, enclos): is_positive :
#    customer_data$Id contains non-positive values.
# There was 1 failure:
#   Position Value  Cause
# 1       14   -14 too low

assert_all_are_whole_numbers(customer_data$Id)
assert_has_no_duplicates(customer_data$Id)

# Error in eval(expr, envir, enclos): has_no_duplicates :
#    customer_data$Id has a duplicate at position 16.
```

Checking the "Title" field for a character vector of honorifics.

```
library(assertive.data)
assert_is_character(customer_data$Title)
assert_all_are_honorifics(customer_data$Title)

# Error in eval(expr, envir, enclos): is_honorific :
```

```
#   customer_data$Title are not all honorifics.
# There were 7 failures:
#  Position     Value        Cause
# 1          3              bad format
# 2          4         Prof bad format
# 3          7        Ninja bad format
# 4         12          Mrr bad format
# 5         14 Brigadier bad format
# 6         19       Rt Hon bad format
# 7         20        Darth bad format
```

Checking the "FirstName" field for a character vector of non-empty, non-missing strings.

```
library(assertive.strings)
assert_is_character(customer_data$FirstName)
assert_all_are_non_missing_nor_empty_character(
  customer_data$FirstName
)

# Error in eval(expr, envir, enclos):
#   is_non_missing_nor_empty_character : customer_data$FirstName
#   are not all non-missing nor non-empty strings.
# There was 1 failure:
#  Position Value Cause
# 1        20     empty
```

Checking the "LastName" field for a character vector of strings of a sensible length. Sensible depends upon the demographic of your customers – Chinese surnames are typically shorter than Sri Lankan ones, for example. In general, it's best to err on the side of treating the data as OK, so a range of two to thirty characters seems reasonable.

```
assert_is_character(customer_data$LastName)
nch_last_name <- nchar(customer_data$LastName)
assert_all_are_in_closed_range(nch_last_name, 2, 30)

# Error in eval(expr, envir, enclos): is_in_closed_range :
#   nch_last_name are not all in the range [2,30].
# There was 1 failure:
#  Position Value  Cause
# 1        1      1 too low
```

Checking the "DateOfBirth" field for dates in the form %Y-%m-%d, implying ages between eighteen and one hundred.

Parsing dates and times can give unexpected results. Notice that the badly formatted date "06-05-1957" still parses to a real date (though not the right one!). This is why the extra check for a plausible age is necessary.

You can get a slightly more accurate age using **periods** from the *lubridate* [14] package, but for these purposes, base-R's **difftime** is OK.

```
library(assertive.datetimes)
assert_all_are_date_strings(
  customer_data$DateOfBirth,
  "%Y-%m-%d"
)

# Error in eval(expr, envir, enclos): is_date_string :
#   customer_data$DateOfBirth contains strings that are not in
#   the date format '%Y-%m-%d'.
# There was 1 failure:
#  Position     Value        Cause
# 1        10 1964-13-15 bad format

customer_data$DateOfBirth <- as.Date(
  customer_data$DateOfBirth,
  "%Y-%m-%d"
)
age <- difftime(
  Sys.Date(),
  customer_data$DateOfBirth,
  units = "days"
) / 365.25
assert_all_are_in_range(age, 18, 100)

# Warning: Coercing age to class 'numeric'.
# Error in eval(expr, envir, enclos): is_in_range :    age are not all in
the range [18,100].
# There were 4 failures:
#  Position              Value    Cause
# 1          6    2010.54346338125  too high
# 2          8   -62.0643394934976  too low
# 3         10                 <NA>  missing
# 4         18   16.5284052019165  too low
```

Checking the "Telephone" field for a character vector of UK telephone numbers.

```
library(assertive.data.uk)
assert_is_character(customer_data$Telephone)
assert_all_are_uk_telephone_numbers(
  customer_data$Telephone
)

# Error in eval(expr, envir, enclos): is_uk_telephone_number :
#   customer_data$Telephone are not all UK telephone numbers.
# There were 11 failures (showing the first 10):
#     Position       Value       Cause
```

```
#  1        2 011422971370 bad format
#  2        7              bad format
#  3        8   02093078556 bad format
#  4       10              bad format
#  5       12  0125255354 bad format
#  6       13              bad format
#  7       15              bad format
#  8       16  1617090166 bad format
#  9       17  01135676514 bad format
# 10    19       9802575 bad format
```

Checking the "Postcode" field for a character vector of UK postcodes.

```
assert_is_character(customer_data$Postcode)
assert_all_are_uk_postcodes(customer_data$Postcode)

# Error in eval(expr, envir, enclos): is_uk_postcode :
#   customer_data$Postcode are not all UK postcodes.
# There were 19 failures (showing the first 10):
#    Position      Value       Cause
# 1          1  KA12 8SE   bad format
# 2          2    S7 1FF   bad format
# 3          3              bad format
# 4          4  CH64 4DU   bad format
# 5          5              bad format
# 6          6  N10 1QX    bad format
# 7          7  GU12 6DD   bad format
# 8          8  W4 3QYQ    bad format
# 9          9    B650HE   bad format
# 10        11 CH100 9AL   bad format
```

A.2.7   Exercise: Calculating the Harmonic Mean

You can more or less copy and paste from the geometric mean example.

```
harmmean <- function(x, na.rm = FALSE)
{
  assert_is_numeric(x)
  if(any(x == 0, na.rm = TRUE))
  {
    warning("The harmonic mean requires non-zero inputs.")
    return(NaN)
  }
  na.rm <- coerce_to(use_first(na.rm), "logical")
  1 / mean(1 / x, na.rm = na.rm)
}
```

Your set of tests to ensure that the function works may be different from the ones here. Are there any tests that you missed? Are there any tests that you thought of that aren't included here?

Let's start with a sensible input: some non-zero numbers. This should return a single numeric value.

```
harmmean(rnorm(20))
```

```
# [1] -15.23954
```

Now let's try including a zero. This should return NaN, with a warning.

```
harmmean(-1:1)
```

```
# Warning in harmmean(-1:1):
#   The harmonic mean requires non-zero inputs.
```

```
# [1] NaN
```

Let's test a non-numeric input. This should throw an error.

```
harmmean("1.234")
```

```
# Error in harmmean("1.234"): is_numeric : x is not of class
#   'numeric'; it has class 'character'.
```

Now let's test the `na.rm` argument. With an `x` input with some missing values, the return value should be `NA` when `na.rm` is `FALSE` and a number when `na.rm` is `TRUE`.

```
x <- rnorm(20)
x[c(2, 5, 11, 17)] <- NA
harmmean(x)
```

```
# [1] NA
```

```
harmmean(x, na.rm = TRUE)
```

```
# [1] -0.6802961
```

Finally, let's test a bad `na.rm` input. This should warn about the problems, and try to correct the input.

```
harmmean(x, na.rm = 1:10)
```

```
# Warning: Only the first value of na.rm (= 1) will be used.
# Warning: Coercing use_first(na.rm) to class 'logical'.
```

```
# [1] -0.6802961
```

A.3   Chapter 3

A.3.1   Exercise: Using `expect_equal`

The test for a 5-12-13 triangle follows the same pattern as the 3-4-5 example in the text.

```
test_that(
  "hypotenuse, with inputs x = 5 and y = 12, returns 13",
  {
    expected <- 13
    actual <- hypotenuse(5, 12)
    expect_equal(actual, expected)
  }
)
```

To make the test fail, you need to change the expected value by about 2e-7 or more. For example, `expected <- 13 + 2e-7`.

Now let's try testing large inputs, close to the largest numbers that can be represented in numeric vectors.

```
test_that(
  "hypotenuse, with inputs x = 1e300 and y = 1e300,
  returns 1.414e300",
  {
    expected <- sqrt(2) * 1e300
    actual <- hypotenuse(1e300, 1e300)
    expect_equal(actual, expected)
  }
)
```

```
# Error: Test failed: 'hypotenuse, with inputs x = 1e300 and y = 1e300,
#   returns 1.414e300'
# * 'actual' not equal to 'expected'.
# 1/1 mismatches
# [1] Inf - 1.41e+300 == Inf
```

The test fails because when you square `1e300`, the value overflows to infinity. This naive algorithm for calculating hypotenuses is pretty bad numerically, and you shouldn't use it in real applications.

For small inputs, we have the same problem: when you square `1e-300`, the value underflows to zero.

```
test_that(
  "hypotenuse, with inputs x = 1e-300 and y = 1e-300,
  returns 1.414e-300",
  {
```

```
    expected <- sqrt(2) * 1e-300
    actual <- hypotenuse(1e-300, 1e-300)
    expect_equal(actual, expected)
  }
)
```

Unfortunately, `expect_equal` won't spot the problem. Take a look in the main text for how to fix this.

A.3.2 Exercise: Using `expect_error`

With a non-numeric input, hypotenuse should throw an error. Notice that when we match the error message, the second argument to `expect_error` is a regular expression, so the hyphen needs to be escaped.

```
test_that(
  "hypotenuse, with input x = '1' throws an error",
  {
    expect_error(
      hypotenuse("1", 1),
      "non\\-numeric argument to binary operator"
    )
  }
)
```

A.3.3 Exercise: Using `expect_output`

The test uses both `expect_output` and `expect_equal`. The line of output is just the numbers one to five separated by spaces, and the return value is the same as the input x.

```
test_that(
  "print, with input x = 1:5 displays output and returns x",
  {
    x <- 1:5
    expect_output(actual <- print(x), "1 2 3 4 5")
    expect_equal(actual, x)
  }
)
```

A.3.4 Exercise: Find More Tests for `square_root`

We need to test

- That numeric vectors with length more than one correctly return square roots for each input number. At the moment this won't work due to that pesky `if` statement, so a call to `Vectorize` or `vapply` will need to be added to the function.

- What should happen with complex inputs? Presumably, you want a complex output too.

- That non-numeric inputs fail gracefully. For example, character vector or list inputs should throw an error.

- That multi-dimensional objects behave as you expect. There isn't an obvious answer to what should happen here. The base-R `sqrt` function preserves the dimensions of matrix inputs, and automatically loops over columns of a data frame input. The latter behaviour in particular is very useful for interactive use, but probably too clever for its own good in a programmatic situation. You may decide that you like this behaviour, or you want something different. Either way, you need some tests to clarify what should happen.

- Is the square root of infinity still infinity?

- What happens if you set the tolerance to zero? (Or very small.) Does the algorithm still converge?

Have you thought of any tests that aren't on the list?

A.3.5   Exercise: Testing the Return Type of Replicates

Let's try the simplest chunk of code possible for our test: in each replicate, we just return the value zero. With the example of `sapply`, we needed to test for non-zero and zero length inputs. Here, similarly, we need to test for non-zero and zero replicates.

```
test_that(
  "replicate with more than zero reps simplifies to a vector",
  {
    actual <- replicate(5, 0)
    expect_type(actual, "double")
  }
)


test_that(
  "replicate with zero reps returns a list",
  {
    actual <- replicate(0, 0)
    expect_type(actual, "list")
  }
)
```

For completeness, you may also want to test non-zero replicates with the argument `simplify = FALSE`.

```
test_that(
  "replicate with more than zero reps, simplify=FALSE
  returns a list",
  {
    actual <- replicate(5, 0, simplify = FALSE)
    expect_type(actual, "list")
  }
)
```

A.4   Chapter 4

A.4.1   Exercise: Reducing duplication

To eliminate the duplicated code, you should adopt a "split-apply-combine" strategy. That is, split the values (weights) into groups (by Diet), apply a summary function (mean), and combine the results back into a single variable.

A common base R solution using `tapply` requires some help from `with` to access columns in the data frame, and it returns the result as a named vector rather than a data frame, which is usually less convenient to work with.

```
with(
  ChickWeight,
  tapply(weight, Diet, mean)
)

#        1        2        3        4
# 102.6455 122.6167 142.9500 135.2627
```

This can be made slightly prettier using the dollar operator in *magrittr*. Here is the same thing again.

```
ChickWeight %$% tapply(weight, Diet, mean)

#        1        2        3        4
# 102.6455 122.6167 142.9500 135.2627
```

I rather like the *dplyr* approach, since the code is very explicit about what is happening. This version uses the standard-evaluation versions, which are best for programming with.

```
library(dplyr)
ChickWeight %>%
  group_by_(~ Diet) %>%
  summarise_(MeanWeight = ~ mean(weight))

# # A tibble: 4  2
#      Diet MeanWeight
#    <fctr>      <dbl>
# 1       1   102.6455
# 2       2   122.6167
# 3       3   142.9500
# 4       4   135.2627
```

The *data.table* solution is more compact, though it requires converting the dataset into a data table and uses some not-very-self-explanatory commands like ..

```
library(data.table)
as.data.table(ChickWeight)[
  j  = .(MeanWeight = mean(weight)),
  by = .(Diet)
]

#     Diet MeanWeight
# 1:     1   102.6455
# 2:     2   122.6167
# 3:     3   142.9500
# 4:     4   135.2627
```

The *plyr* solution is a little more complicated.

```
library(plyr)
ddply(
  ChickWeight,
  .(Diet),
  summarize,
  MeanWeight = mean(weight)
)

#   Diet MeanWeight
# 1    1   102.6455
# 2    2   122.6167
# 3    3   142.9500
# 4    4   135.2627
```

A.4.2   Exercise: Outsourcing Argument Checking

`trainControl` successfully stops nonsensical inputs to some of the arguments, but not others. For example, `search` is checked:

```
library(caret)
trainControl(search = "nonsense")

# Error in trainControl(search = "nonsense"): 'search' should be either
#    'grid' or 'random'
```

...but not `method`. Only the first few terms are shown for brevity.

```
head(trainControl(method = "nonsense"))

# $method
# [1] "nonsense"
#
# $number
# [1] 25
#
# $repeats
# [1] 25
#
# $search
# [1] "grid"
#
# $p
# [1] 0.75
#
# $initialWindow
# NULL
```

A good update to `trainControl` would be to use `match.arg` on all the string arguments that must take specific values.

```
trainControl <- function(method = c("boot", "boot632",
  "cv", "repeatedcv", "LOOCV", "LGOCV", "none", "oob",
  "adaptive_cv", "adaptive_boot", "adaptive_LGOCV"), ...)
  # plus the other args
{
  method <- match.arg(method)
  # rest of the function
}
```

A test for this might be

```
test_that(
  "trainControl, with a nonsense method, throws an error",
  {
    expect_error(
      trainControl(method = "nonsense"),
      "'arg' should be one of"
    )
  }
)
```

Other useful improvements include making sure that arguments like **number** and **repeats** are positive integers, and that the training proportion, **p**, is between zero and one.

A.4.3 Exercise: Wrappers for Formatting Currency

All you need is **paste0** and **formatC**.

```
format_as_usd <- function(x)
{
  paste0("$", formatC(x, digits = 2, format = "f"))
}
# Usage:
format_as_usd(c(1, 1.23e6, 1.2345))

# [1] "$1.00"     "$1230000.00" "$1.23"
```

A.4.4 Exercise: Providing Better Defaults for write.csv

The first two updates are straightforward – we just need to set **row.names = FALSE** and **na = ""**. Autogenerating a file name is slightly trickier because the name needs to be a valid file name. We need to replace some punctuation characters like slashes with a safe character, like an underscore.

In the example below, the code for autogenerating the name is in the body of the function. You can keep it in the function signature if you prefer, but since it's a little long and complicated, it is harder to read there. Note the use of *magrittr*'s piping to make the code clearer, and *stringi*'s regular-expression replacement of the punctuation.

```
library(assertive.base)
library(magrittr)
library(stringi)
write_csv <- function(x, file = NULL, row.names = FALSE,
```

```
  na = "", ...)
{
  if(is.null(file))
  {
    file <- get_name_in_parent(x) %>%
      stri_replace_all_regex("[\\\\/:*?\"<>|]", "_") %>%
      paste0(".csv")
    message("Writing to ", file)
  }
  write.csv(x, file = file, na = na, row.names = row.names, ...)
}
```

A.4.5   Exercise: Decomposing the `quantile` Function

The first obvious simplification to `quantile.default` is that a lot of
the custom argument checking can be handled by *assertive*. For example,
the check that the `prob` argument is between 0 and 1 can be handled
by `assert_all_are_proportions`. Beyond that, the most obvious mess in
`quantile.default` is caused by implementing nine different algorithms! Actually, the code for `types` 1 to 3, used on discrete inputs, is all quite similar,
and likewise the code for `types` 4 to 9 is also quite similar. It makes sense
to outsource these for two different functions. You could write nine separate
backend functions for each of the algorithms. This is perfectly valid, but you'd
have to be careful about not duplicating code.

There are also bits of code for dealing with missing values, zero
length inputs, and factors. For factor input, the code doesn't belong in
`quantile.default` at all. This ought to be in a `quantile.factor` method.

A rough implementation would look something like

```
quantile.default <- function (x, probs = seq(0, 1, 0.25),
  na.rm = FALSE, names = TRUE, type = 7, ...)
{
  # Check/fix the inputs
  assert_all_are_proportions(
    probs,
    tol = 100 * .Machine$double.eps
  )
  x <- handle_nas(x, na.rm)
  # ... and other checks

  # Deal with edge case of zero length inputs
  if(is_empty(x) || is_empty(probs))
  {
    return(handle_empty_inputs(x, probs))
  }
```

```
#Choose a quantile algorithm function
quantile_fn <- if(type %in% 1:3)
{
  quantile_discrete
} else {
  quantile_continuous
}
quantile_fn(x, probs, type)
}
```

A.4.6 Exercise: Calculating Leap Years

It's best to completely remove the `if` statements and use R's built-in vectorization of its operators.

```
is_leap_year2 <- function(year)
{
  yn <- year %% 4 == 0
  yn[year %% 100 == 0] <- FALSE
  yn[year %% 400 == 0] <- TRUE
  setNames(yn, year)
}
```

A slightly fancier variation is to use `is_divisible_by`, from *assertive.numbers*.

```
library(assertive.numbers)
is_leap_year3 <- function(year)
{
  yn <- is_divisible_by(year, 4)
  yn[is_divisible_by(year, 100)] <- FALSE
  yn[is_divisible_by(year, 400)] <- TRUE
  setNames(yn, year)
}
```

A.5 Chapter 5

A.5.1 Exercise: Make a Package with Tests

First we create the package and its description file.

```
library(devtools)
descriptionDetails <- list(
  Title       = "Calculate Square Roots",
  Version     = "0.0-1",
  Author      = "Arthur Power [aut,cre]",
  Maintainer  = "Arthur Power <a@p.com>",
  Description = "A square root fn, plus tests!",
  Licence     = "GPL-3",
  URL         = "https://package-homepage.com",
  BugReports  = "https://package-homepage.com/issues"
)
create("rooty", description = descriptionDetails)

# Creating package 'rooty' in
#   '/Users/richierocks/workspace/testingrcode/chapters/appendices'
# No DESCRIPTION found. Creating with values:
# * Creating 'rooty.Rproj' from template.
# * Adding '.Rproj.user', '.Rhistory', '.RData' to ./.gitignore
```

Next we write the contents of the R file.

```
writeLines(
  "#' Calculate hypotenuses
#'
#' Calculates the square root, using the Babylonian method.
#' @param x A number.
#' @param tol A positive number. How close does the answer need to be?
#' @return The square root.
#' @export
square_root_v3 <- function(x, tol = 1e-6)
{
  if(is_empty(x))
  {
    return(numeric())
  }
  if(is_negative(x))
  {
    warning('Negative inputs are not supported; returning NaN.')
    return(NaN)
  }
  S <- x
  x <- log2(x) ^ 2
  repeat{
    x <- 0.5 * (x + (S / x))
    err <- x ^ 2 - S
    if(abs(err) < tol)
    {
      break
    }
  }
  x
}",
  con = "rooty/R/square-root.R"
)
```

Next we add testing infrastructure.

```
use_testthat("rooty")

# * Adding testthat to Suggests
# * Creating 'tests/testthat'.
# * Creating 'tests/testthat.R' from template.
```

Then we can add the tests.

```
writeLines(
  'context("Testing the square_root_v3 function")
test_that(
  "square_root_v3, with input 1024, returns 32",
  {
    expected <- 32
    actual <- square_root_v3(1024)
    expect_equal(actual, expected)
  }
)
test_that(
  "square_root_v3, with a negative input,
  returns NaN with a warning",
  {
    expected <- NaN
    expect_warning(
      actual <- square_root_v3(-1),
      "Negative inputs are not supported; returning NaN."
    )
    expect_equal(actual, expected)
  }
)
test_that(
  "square_root_v3, with a zero-length numeric input,
  returns a zero-length numeric",
  {
    expected <- numeric()
    actual <- square_root_v3(numeric())
    expect_equal(actual, expected)
  }
)',    # etc.
  con = "rooty/tests/testthat/test-square-root.R"
)
```

To make sure this works, we need to build the package. (You should really check it too.)

```
build("rooty")
```

```
# '/Library/Frameworks/R.framework/Resources/bin/R' --no-site-file --no-environ \
#  --no-save --no-restore --quiet CMD build \
```

```
#  '/Users/richierocks/workspace/testingrcode/chapters/appendices/rooty' \
#  --no-resave-data --no-manual
#
```

A.6 Chapter 6

A.6.1 Exercise: Writing a Test That Handles Side Effects

We use `with_options` to temporarily change the global `digits` option.

```
test_that(
  "Printing pi to 10 digits outputs 3.141592654",
  {
    expected <- "3.141592654"
    with_options(
      c(digits = 10),
      expect_output(print(pi), expected)
    )
  }
)
```

A.6.2 Exercise: Testing a Complex Object

1. To check that the extra hours of sleep are significantly different between the groups, we just need an assertion.

```
library(assertive.numbers)
assert_all_are_less_than(test_results$p.value, 0.05)
```

2. For testing the structure of the object, the list of possible tests is almost endless. Here, as a good start, I think it's worth testing that the results have all the elements appropriate to an object of class `htest`, that the p-value is less than 0.05, and the degrees of freedom are 9 (one less than the sample size of each group).

```
test_that(
  "the elements of the model are correct",
  {
    expected <- c(
      "statistic", "parameter", "p.value",
      "conf.int", "estimate", "null.value",
      "alternative", "method", "data.name"
```

```
    )
    actual <- names(test_results)
    expect_equal(actual, expected)
  }
)


test_that(
  "the p-value is less than 0.05",
  {
    actual <- test_results$p.value
    expect_less_than(actual, 0.05)
  }
)


test_that(
  "the p-value is less than 0.05",
  {
    expected <- c(df = 10 - 1)
    actual <- test_results$parameter
    expect_equal(actual, expected)
  }
)
```

A.6.3   Exercise: Testing an *Rcpp* Function

First, we create a package.

```
descriptionDetails <- list(
  Title = "Calculate squares",
  Version = "0.0-1",
  "Authors@R" =
    'person("Tu", "Powa", email = "t@p.com", role = c("aut", "cre"))',
  Description =
  "A square fn, plus tests!",
  License = "GPL-3",
  URL = "https://package-homepage.com",
  BugReports = "https://package-homepage.com/bug-tracker",
  Imports = "Rcpp",
  Suggests = "testthat",
  LinkingTo = "Rcpp, testthat"
)
pkg <- "square"
create(pkg, description = descriptionDetails)

# Error: Directory exists and is not empty
```

Add the infrastructure.

```
use_rcpp(pkg)

# Adding Rcpp to LinkingTo and Imports
# * Ignoring generated binary files.
```

```
# Next, include the following roxygen tags somewhere in your package:
#
# #' @useDynLib square
# #' @importFrom Rcpp sourceCpp
# NULL
#
# Then run document()
```

```
use_testthat(pkg)
```

```
# * testthat is already initialized
```

```
use_catch(pkg)
```

```
# > Added C++ unit testing infrastructure.
# > Please ensure you have 'LinkingTo: testthat' in your DESCRIPTION.
# > Please ensure you have 'useDynLib(square)' in your NAMESPACE.
```

```
writeLines(
"#' @useDynLib square
#' @importFrom Rcpp sourceCpp
NULL",
    file.path(pkg, "R/rcpp-setup.R")
)
```

Add the function, with Catch tests.

```
writeLines(
  '#include <Rcpp.h>
using namespace Rcpp;

// [[Rcpp::export]]
NumericVector square(NumericVector x) {
  return x * x;
}

context("test square") {
  test_that("the square fn works") {
    NumericVector x = NumericVector::create(2.5, 71.0);
    NumericVector expected = NumericVector::create(6.25, 5041.0);
    NumericVector actual = square(x);
    expect_true(is_true(all(actual == expected)));
  }
}',
  file.path(pkg, "src/square.cpp")
)
```

Write the documentation for the function.

```
writeLines(
  "#' Calculate two sine.
#'
#' Calculates the sine of twice the input, using C++.
#' @param x A numeric vector.
#' @return Two sine.
```

```
#' @name square
#' @export
NULL",
  file.path(pkg, "R/square.R")
)
```

Generate documentation files, build, and check.

```
roxygenize(pkg)
build(pkg)
check(pkg)
```

A.6.4   Exercise: Writing INI Configuration Files

To write the INI file, it's just a bit of fiddling with `paste` to construct the lines of output and `cat` to write it to the file or connection. This implementation writes an extra blank line, but the extra fiddling to avoid that isn't worth it for this exercise.

```
write_ini <- function(x, file)
{
  lines <- Map(
    function(section, elements)
    {
      c(
        paste0("[", section, "]"),
        paste(names(elements), elements, sep = "="),
        ""
      )
    },
    names(x),
    x
  )
  cat(unlist(lines), file, sep = "\n")
}
```

To test it, we write to the console (a.k.a. the standard output connection) and use `capture.output`. Since the function uses `cat` to do the writing, we use the special syntax of `file = ""` to write to the console, rather than using `stdout()`.

```
test_that(
  "write_ini correctly writes to stdout",
  {
    ini <- list(
      section1 = c(element1 = 1, element2 = "a"),
```

```
      section2 = c(element3 = TRUE)
   )
   expected <- c(
     "[section1]",
     "element1=1",
     "element2=a",
     "",
     "[section2]",
     "element3=TRUE"
   )
   # Use "" to write to stdout
   actual <- capture.output(write_ini(ini, ""))
   # Don't worry about trailing "" lines
   actual <- actual[1:max(which(nzchar(actual)))]
   expect_equal(actual, expected)
  }
)
```

A.6.5   Exercise: Write a Graphics Test Report

This report follows a similar format to the one in the chapter. It's written in markdown, but feel free to try and translate it into LaTeX or Asciidoc or one of the other supported languages.

A.7   Chapter 7

A.7.1   Exercise: Writing a Custom Scalar Predicate

The function needs two inputs: x, for the vector to test, and .xname for the name of that input in the parent environment. For the input to be monotonically increasing, diff(x) should be non-negative everywhere. We test for the converse situation, that is, if diff(x) is negative anywhere, we return FALSE.

```
is_monotonic_increasing <- function(x,
   .xname = get_name_in_parent())
{
  if(any(diff(x) < 0))
  {
    return(
      false("%s is not monotonically increasing.", .xname)
    )
  }
  TRUE
}
```

You could be even more helpful to the user, and – in the event of failure – return the positions and values where things go wrong.

It's a good idea to include values that are equal in at least one test, since this is a potential area of confusion. (Where you have consecutive equal values, a vector can be monotonically increasing, but not *strictly* monotonically increasing.)

```
test_that(
  paste(
    "is_monotonic_increasing returns TRUE",
    "for a monotonically increasing input"
  ),
  {
    actual <- is_monotonic_increasing(c(-1, 3, 3, 6.6))
    expect_true(actual)
  }
)
test_that(
  paste(
    "is_monotonic_increasing returns FALSE",
    "for a non-monotonically increasing input"
  ),
  {
    actual <- is_monotonic_increasing(c(-1, 3, 2, 6.6))
    expect_false(actual)
    expect_match(
      cause(actual),
      "is not monotonically increasing"
    )
  }
)
```

A.7.2 Exercise: Writing a Custom Scalar Assertion

This function should just call assert_engine with two inputs: x, the input to be checked, and severity, to measure the consequences of failure.

```
assert_is_monotonic_increasing <- function(x,
  severity = getOption("assertive.severity", "stop"))
{
  assert_engine(
    is_monotonic_increasing,
    x,
    .xname = get_name_in_parent(x),
    severity = severity
```

```
  )
}
```

A.7.3   Exercise: Writing a Vector Predicate

The function just needs a single argument, x. We call `grepl` to match the DNA sequence, then add a cause attribute and names using `set_cause` and `call_and_name`, respectively.

```
is_dna_seq <- function(x)
{
  dna_regex <- "^[ACGT]+$"
  x <- coerce_to(x, "character")
  call_and_name(
    function(x)
    {
      ok <- grepl(dna_regex, x, ignore.case = TRUE)
      set_cause(ok, "bad format")
    }
  )
}
```

A.7.4   Exercise: Writing a Vector Assertion

We need two lines: one to create a message and one to call `assert_engine`.

```
assert_all_are_dna_seqs <- function(x, na_ignore = FALSE,
  severity = getOption("assertive.severity", "stop"))
{
  msg <- gettextf(
    "%s are not all DNA sequences.",
    get_name_in_parent(x)
  )
  assert_engine(
    is_dna_seq,
    x,
    na_ignore = na_ignore,
    severity = severity
  )
}
```

The "any" version just tweaks the failure message and adds the additional line `what = "any"`.

```
assert_any_are_dna_seqs <- function(x, na_ignore = FALSE,
  severity = getOption("assertive.severity", "stop"))
{
  msg <- gettextf(
    "%s are all not DNA sequences.",
    get_name_in_parent(x)
  )
  assert_engine(
    is_dna_seq,
    x,
    what = "any",
    na_ignore = na_ignore,
    severity = severity
  )
}
```

A.7.5   Exercise: Create a Custom Expectation

The expectation contains the same boilerplate code as the example in the text; only the condition for success and the failure message have changed.

```
expect_positive <- function(object, info = NULL, label = NULL)
{
  lab <- testthat:::make_label(object, label)
  expect(
    is.numeric(object) && all(object > 0),
    sprintf(
      "%s is not a numeric vector of positive values.",
      lab
    ),
    info = info
  )
  invisible(object)
}
```

To check that it works, here are some tests.

```
test_that(
  "a vector of positive numbers passes the test",
  {
    expect_positive(rlnorm(10))
  }
)


test_that(
  "a vector with some negative numbers fails the test",
  {
```

```
    expect_positive(-5:5)
  }
)
```

```
# Error: Test failed: 'a vector with some negative numbers fails the test'
# * -5:5 is not a numeric vector of positive values.
```

Since having a failing test is unsatisfying, the meta-test equivalent is

```
test_that(
  "expect_positive fails when there are negative inputs",
  {
    expect_error(
      test_that(
        "a vector with some negative numbers fails the test",
        {
          expect_positive(-5:5)
        }
      )
    )
  }
)
```

Bibliography

[1] Stefan Milton Bache and Hadley Wickham. *magrittr: A Forward-Pipe Operator for R*, 2014. R package version 1.5.

[2] Winston Chang. *vtest: Tools for visual testing of R packages*. R package version 0.0.3.

[3] Joe Conway, Dirk Eddelbuettel, Tomoaki Nishiyama, Sameer Kumar Prayaga, and Neil Tiffin. *RPostgreSQL: R interface to the PostgreSQL database system*, 2013. R package version 0.4.

[4] Richard Cotton. *Learning R*. O'Reilly, 2013.

[5] Richard Cotton. *rebus.datetimes: Date and Time Extensions for the 'rebus' Package*, 2015. R package version 0.0-1.

[6] Richard Cotton. *runittotestthat: Convert 'RUnit' Test Functions into 'testthat' Tests*, 2015. R package version 0.0-2.

[7] Richard Cotton. *sig: Print Function Signatures*, 2015. R package version 0.0-5.

[8] Richard Cotton. *assertive: Readable Check Functions to Ensure Code Integrity*, 2016. R package version 0.3-3.

[9] Gabor Csardi. *cyclocomp: Cyclomatic Complexity of R Code*, 2016. R package version 1.0.0.

[10] Dirk Eddelbuettel. *Seamless R and C++ Integration with Rcpp*. Springer, 2013.

[11] Max Kuhn. Contributions from Jed Wing, Steve Weston, Andre Williams, Chris Keefer, Allan Engelhardt, Tony Cooper, Zachary Mayer, Brenton Kenkel, the R Core Team, Michael Benesty, Reynald Lescarbeau, Andrew Ziem, Luca Scrucca, Yuan Tang, and Can Candan. *caret: Classification and Regression Training*, 2016. R package version 6.0-68.

[12] Marek Gagolewski and Bartek Tartanus. *R package stringi: Character string processing facilities*, 2015.

[13] Tal Galili. *installr: Using R to Install Stuff (Such As: R, Rtools, RStudio, Git, and More!)*, 2016. R package version 0.17.8.

[14] Garrett Grolemund and Hadley Wickham. Dates and times made easy with lubridate. *Journal of Statistical Software*, 40(3):1–25, 2011.

[15] Jim Hester. *covr: Test Coverage for Packages*, 2015. R package version 1.2.0.

[16] Jim Hester, Kirill Müller, Hadley Wickham, and Winston Chang. *withr: Run Code 'With' Temporarily Modified Global State*, 2016. R package version 1.0.2.

[17] R Special Interest Group on Databases. *DBI: R Database Interface*, 2014. R package version 0.3.1.

[18] Bioconductor Package Maintainer. *BiocCheck: Bioconductor-specific package checks*. R package version 1.6.1.

[19] Inc Stack Exchange. Stack overflow, 2016.

[20] R Core Team. R-package-devel mailing list, 2016.

[21] Hadley Wickham. *ggplot2: Elegant Graphics for Data Analysis*. Springer-Verlag New York, 2009.

[22] Hadley Wickham. The split-apply-combine strategy for data analysis. *Journal of Statistical Software*, 40(1):1–29, 2011.

[23] Hadley Wickham. testthat: Get started with testing. *The R Journal*, 3:5–10, 2011.

[24] Hadley Wickham. *Advanced R*. CRC Press, 2014.

[25] Hadley Wickham. *R Packages*. O'Reilly, 2015.

[26] Hadley Wickham. *stringr: Simple, Consistent Wrappers for Common String Operations*, 2015. R package version 1.0.0.

[27] Hadley Wickham and Winston Chang. *devtools: Tools to Make Developing R Packages Easier*, 2016. R package version 1.10.0.

[28] Hadley Wickham, Peter Danenberg, and Manuel Eugster. *roxygen2: In-Source Documentation for R*, 2015. R package version 5.0.1.

[29] Hadley Wickham and Romain Francois. *dplyr: A Grammar of Data Manipulation*, 2015. R package version 0.4.3.

[30] Hadley Wickham and Jeroen Ooms. *RPostgres: Experimental Rcpp Interface to PostgreSQL*, 2016. R package version 0.1-2.

[31] Dirk Eddelbuettel with contributions by Antoine Lucas, Jarek Tuszynski, Henrik Bengtsson, Simon Urbanek, Mario Frasca, Bryan Lewis, Murray Stokely, Hannes Muehleisen, Duncan Murdoch, Jim Hester, Wush Wu, and Thierry Onkelinx. *digest: Create Compact Hash Digests of R Objects*, 2016. R package version 0.6.9.

[32] Yihui Xie. *Dynamic Documents with R and knitr, Second Edition*. CRC Press, 2015.

Concept Index

assertions, *see* run-time testing
atomic variables, 16, 22, 109

C++, 39, 50, 78, 102–107, 115, 148
Catch C++ unit testing framework, 103
checking packages, 84, 85
    R CMD check, 82
cleaning data, 26
continuous integration, 54, 86–89
    SemaphoreCI, 85
    Travis CI, 85, 88
Cyclomatic complexity, 73
cyclomatic complexity, 73–76

database
    PostgreSQL, 96
dates and times, 1–3, 6, 8, 10, 24, 25, 32, 35, 42, 46, 50, 51, 53, 57, 62, 63,
    76, 79, 85, 86, 88, 96, 99, 100, 109, 117, 131, 132
debugging, 28
DESCRIPTION, *see* package contents, *see* package contents, *see* package
    contents, *see* package contents
development-time testing, xiv, 2, 35, 54, 83, 91

examples, 82

fail early, fail often, 29

gcc command line tool, 102
global options, 63, 91, 92, 120, 146
graphics, xiv, 24, 69, 110, 114, 115

IDE, xv, 23, 24, 77, 79, 81–84, 114
integrated development environment, xv

LICENCE, *see* package contents
LICENSE, *see* package contents

make command line tool, 102

man page, *see* help page

NAMESPACE, *see* package contents, *see* package contents
NEWS, *see* package contents

recursive variables, 14, 16
report, 53, 54, 58, 62, 66, 67, 111, 114, 115, 150
run-time testing, xiv, 2, 7, 12, 32, 35

scatter plot, 112, 114
side effect, 7, 91, 92, 115
src, *see* package contents, *see* package contents, *see* package contents, *see*
          package contents

unit testing, *see* development-time testing

Package Index

assertive, xiv, xv, 2, 7, 11, 29,
        31–33, 67, 117, 124, 142
assertive.code, 27
assertive.data, 25
assertive.data.uk, 25
assertive.data.us, 25
assertive.datetimes, 24
assertive.files, 19, 129
assertive.matrices, 21
assertive.models, 22
assertive.numbers, 12, 17, 21, 45,
        121, 130, 143
assertive.properties, 12, 13, 16,
        118
assertive.reflection, 23, 86
assertive.sets, 22
assertive.strings, 20
assertive.types, 12, 16, 119

BiocCheck, xv, 83

caret, xv, 69
covr, xv, 88
cyclocomp, xv, 74, 76

data.table, xv, 49, 71, 139
DBI, 96
devtools, xv, 78, 79, 82, 84, 85,
        102, 103, 114
digest, xv
dplyr, xv, 72, 96, 100, 138

ggplot2, xv, 58, 63, 64, 69

grid, 68, 69

installr, xvi, 77

knitr, xv–xvii, 8, 110, 114, 115

lubridate, 132

magrittr, xvi, 8, 40, 58, 138, 141

plyr, xvi, 72, 139

Rcpp, xiv, xvi, 102–104
rebus.datetimes, xvi, 50
roxygen2, xvi, 78, 80
RPostgreSQL, xvi, 96
runittotestthat, xvi, 54

sig, xvi, 66, 76
stats, 73
stringi, 141
stringr, 122

testthat, xiv–xvii, 2, 6, 32, 42, 51,
        53, 54, 67, 78, 79, 93, 97,
        103, 104, 106, 114, 115,
        117, 124

utils, 66, 70, 71

vtest, 110

withr, xvi, 92, 115

Dataset Index

ChickWeight, 65
chickwts, 22, 65
customer_data, 26

diamonds, 58

Harman23.cor, 19

longley, 22

stack.x, 14

women, 22

People Index

Bache, Stefan M., 8

Csardi, Gabor, 74, 85

Dowle, Matt, 49

Eddelbuettel, Dirk, 102

Galili, Tal, 77
Gaslam, Brodie, 54
Gillespie, Colin, xvi

Hester, Jim, 88

Kuhn, Max, 69

Srinivasan, Arun, 49

Wickham, Hadley, xiv, xvii, 58, 77, 78, 102

Xie, Yihui, xvii, 110

Function Index

abs (base), 44, 45, 47
add_desc_package (devtools), 103, 105
add_desc_package (devtools, internal), 103
aes (ggplot2), 63, 64, 112
all (base), 7, 153
all.equal (base), 7, 37
any (base), 30, 31, 133, 150
are_same_length (assertive.properties), 15
are_set_equal (assertive.sets), 22
are_set_equal(colnames(women), c(weight, height)) (assertive.sets), 22
as.data.table (data.table), 139
as.Date (base), 132
as.formula (base), 64
as.formula (stats), 64
assert_all_are_date_strings (assertive), 132
assert_all_are_equal_to (assertive.numbers), 129
assert_all_are_finite (assertive.numbers), 128
assert_all_are_honorifics (assertive), 130
assert_all_are_imaginary (assertive.numbers), 122, 123
assert_all_are_in_closed_range (assertive.numbers), 130, 131
assert_all_are_in_range (assertive.numbers), 129, 132
assert_all_are_less_than (assertive.numbers), 146
assert_all_are_nan (assertive.numbers), 28
assert_all_are_non_missing_nor_empty_character (assertive), 131
assert_all_are_non_negative (assertive), 8
assert_all_are_non_negative (assertive.numbers), 8–10, 30
assert_all_are_not_na (assertive.base), 128
assert_all_are_positive (assertive.numbers), 93, 130
assert_all_are_proportions (assertive.numbers), 129, 142
assert_all_are_real (assertive.numbers), 130
assert_all_are_true (assertive.base), 12
assert_all_are_uk_postcodes (assertive.data.uk), 133
assert_all_are_uk_telephone_numbers (assertive), 132
assert_all_are_valid_variable_names (assertive.code), 27
assert_all_are_whole_numbers (assertive), 8
assert_all_are_whole_numbers (assertive.numbers), 8, 130
assert_any_are_imaginary (assertive.numbers), 123

assert_any_are_non_negative (assertive.numbers), 10
assert_any_are_true (assertive), 20
assert_any_are_true (assertive.base), 12
assert_engine (assertive), 117, 118
assert_engine (assertive.base), 117, 120, 123, 151–153
assert_has_colnames (assertive.properties), 129
assert_has_no_duplicates (assertive.properties), 130
assert_has_rownames (assertive.properties), 129
assert_has_rows (assertive.properties), 120
assert_is_character (assertive.types), 130–133
assert_is_list (assertive.types), 129
assert_is_matrix (assertive.types), 129
assert_is_monotonic_increasing (assertive.properties), 151
assert_is_numeric (assertive), 8
assert_is_numeric (assertive.types), 8, 30, 31, 127, 130, 133
assert_is_of_length (assertive.properties), 129
assert_is_r_current (assertive.reflection), 130
assert_is_symmetric_matrix (assertive), 130
assert_r_can_compile_code (assertive), 102
attr (base), 9, 13
attributes (base), 9, 10

base::sqrt (base), 44
break (base), 44, 45, 47
build (devtools), 82, 88, 107, 145, 149

c (base), 4, 5, 7, 10, 12, 13, 15–18, 20–22, 24, 25, 27, 31, 41, 42, 50, 67,
        92–94, 106, 108, 109, 121, 122, 127, 134, 141, 146, 149, 151
call_and_name (assertive), 122
call_and_name (assertive.base), 121, 122, 152
capture.output (utils), 107, 108, 115, 149
cat (base), 107, 149
cause (assertive.base), 151
character (base), 13
check (devtools), 54, 82, 107, 149
class (base), 15
close (base), 108
coefficients (stats), 93, 94, 100
coerce_to (assertive), 31
coerce_to (assertive.base), 133, 152
col (base), 63
context (testthat), 51, 52, 81
cor (stats), 130
cov (stats), 129
create (devtools), 79, 80, 102, 103, 144, 147

cut (base), 58, 63
cyclocomp (cyclocomp), 74, 75

data (utils), 91, 92, 99, 108
data.frame (base), 15, 91, 92, 99, 100, 108
data.table (data.table), 49, 71, 139
date (base), 100
dbConnect (DBI), 96, 98
dbDisconnect (DBI), 97
dbGetQuery (DBI), 96, 97
ddply (plyr), 72, 139
deparse (base), 43
diff (base), 119, 150
difftime (base), 132
digest (digest), 108
dim (base), 13
dir (base), 20
dist (stats), 93, 94

exp (base), 29–31
expect (testthat), 125, 153
expect_cpp_tests_pass (testthat), 39, 50, 103
expect_equal (testthat), 36–42, 44–47, 49, 50, 81, 86, 93–95, 98, 99, 106, 108,
            124, 135, 136, 146, 149
expect_equal_to_reference (testthat), 39, 50, 94, 95
expect_equivalent (testthat), 38, 42, 94
expect_error (testthat), 37–39, 104, 125, 136, 141, 154
expect_false (testthat), 38, 104, 124, 151
expect_gt (testthat), 38, 50
expect_gte (testthat), 50
expect_identical (testthat), 38, 41, 91–93
expect_is (testthat), 38, 49, 93
expect_length (testthat), 38, 49
expect_less_than (testthat), 146
expect_lt (testthat), 38, 50
expect_lte (testthat), 50
expect_match (testthat), 38, 50, 151
expect_message (testthat), 38, 39
expect_named (testthat), 39, 42
expect_not_null, 125
expect_null (testthat), 38, 43
expect_output (testthat), 38, 39, 49, 108, 136, 146
expect_output_file (testthat), 39
expect_positive, 153, 154
expect_s3_class (testthat), 38, 49

expect_s4_class (testthat), 38, 49
expect_silent (testthat), 39, 41, 49
expect_true (testthat), 38, 52, 87, 104, 151
expect_type (testthat), 38, 48, 137, 138
expect_warning (testthat), 38–41, 44, 45, 47

facet_wrap (ggplot2), 63, 64
false (assertive.base), 150
file (base), 141, 149
file.access (base), 129
file.exists (base), 19
file.path (base), 19, 108, 147, 148
filter_ (dplyr), 100
fitted (stats), 94
fitted.values (stats), 94
for (base), 50
format (base), 50
formatC (base), 70, 141
fread (data.table), 49, 71
function (base), 4, 5, 29–32, 35, 44, 45, 47, 64, 74–76, 81, 96–100, 125, 133,
        140–143, 149–153

geom_bar (ggplot2), 63, 64
geom_density (ggplot2), 63
geom_point (ggplot2), 63, 112
geomean, 29
geomean2, 30
geomean3, 31
get_name_in_parent (assertive.base), 141, 151–153
get_revenue_data_from_db, 98–100
gettextf (base), 123, 152, 153
ggplot (ggplot2), 58, 63, 64, 112
gpar (grid), 68, 69
grepl (base), 122, 152
grid (graphics), 68
group_by (dplyr), 72
group_by_ (dplyr), 139
gzfile (base), 108

harmmean, 133, 134
has_any_attributes (assertive), 15
has_any_attributes (assertive.properties), 15
has_attributes (assertive), 15
has_attributes (assertive.properties), 15
has_colnames (assertive.properties), 14
has_cols (assertive.properties), 15

has_dimnames (assertive.properties), 14
has_dims (assertive.properties), 15
has_duplicates (assertive), 16
has_duplicates (assertive.properties), 15
has_elements (assertive.properties), 15
has_names (assertive), 13
has_names (assertive.properties), 13, 14
has_no_attributes (assertive), 15
has_no_attributes (assertive.properties), 15
has_no_duplicates (assertive), 16
has_no_duplicates (assertive.properties), 15
has_rownames (assertive.properties), 14
has_rows (assertive.properties), 15, 118–120
has_terms (assertive.models), 22, 23
have_same_dims (assertive.properties), 15
head (base), 65
head (utils), 140
htest class (stats), 146
hypotenuse, 36, 37, 41–43, 81, 135, 136

identity (base), 109
if (base), 24, 30, 31, 44, 45, 47, 52, 74–76, 133, 141, 142, 150
install.rtools (installr), 77
install_bitbucket (devtools), 84
install_github (devtools), 84
integer vector (base), 119
invisible (base), 125, 153
is.null (base), 124, 125, 141
is.numeric (base), 7, 117, 118, 153
is_a_number (assertive.types), 16
is_a_string (assertive.types), 20
is_after (assertive.datetimes), 24
is_architect (assertive.reflection), 24
is_atomic (assertive.properties), 16
is_batch_mode (assertive.reflection), 24
is_before (assertive.datetimes), 24
is_binding_locked (assertive.code), 27
is_bsd (assertive.reflection), 24
is_character (assertive.types), 119
is_credit_card_number (assertive), 25
is_credit_card_number (assertive.data), 25
is_date_string (assertive), 24
is_date_string (assertive.datetimes), 24
is_diagonal_matrix (assertive.matrices), 21
is_dir (assertive.files), 19

is_divisible_by (assertive.numbers), 143
is_dna_seq, 152, 153
is_empty (assertive), 47
is_empty (assertive.properties), 15, 142
is_empty_character (assertive.strings), 20, 21
is_empty_model (assertive.models), 23
is_equal_to (assertive.numbers), 18
is_executable_file (assertive.files), 19
is_existing_file (assertive.files), 19
is_false (assertive.base), 12
is_file_connection (assertive.files), 20
is_finite (assertive.numbers), 128
is_identical_to_false (assertive.base), 13
is_identical_to_na (assertive.base), 13
is_identical_to_true (assertive), 13
is_identical_to_true (assertive.base), 12
is_identity_matrix (assertive), 21
is_if_condition (assertive.code), 27
is_imaginary (assertive.numbers), 19, 121–123
is_in_future (assertive.datetimes), 25
is_in_past (assertive), 25
is_in_past (assertive.datetimes), 25
is_interactive (assertive.reflection), 24
is_linux (assertive.reflection), 24
is_lower_triangular_matrix (assertive.matrices), 21
is_missing_or_empty_character (assertive.strings), 20, 21
is_monotonic_increasing, 151
is_monotonic_increasing (assertive.properties), 150
is_na (assertive.base), 12
is_negative (assertive.numbers), 17, 30, 31, 45, 47
is_non_empty (assertive.properties), 15
is_non_empty_character (assertive), 20
is_non_empty_character (assertive.strings), 20
is_non_empty_model (assertive.models), 23
is_non_missing_nor_empty_character (assertive), 20
is_non_missing_nor_empty_character (assertive.strings), 20
is_non_negative (assertive.numbers), 10
is_non_scalar (assertive.properties), 15
is_not_equal_to (assertive.numbers), 18
is_not_false (assertive.base), 12
is_not_na (assertive.base), 12, 128
is_not_null (assertive.properties), 16
is_not_true (assertive.base), 12
is_null (assertive.properties), 16
is_numeric (assertive.types), 9, 127

is_numeric_string (assertive.strings), 21
is_of_dimension (assertive.properties), 15
is_of_length (assertive.properties), 15
is_osx (assertive.reflection), 24
is_osx_el_capitan (assertive.reflection), 24
is_proportion (assertive.numbers), 17
is_r (assertive.reflection), 24
is_r_alpha (assertive.reflection), 24
is_r_beta (assertive.reflection), 24
is_r_devel (assertive), 24
is_r_devel (assertive.reflection), 23
is_r_patched (assertive.reflection), 24
is_r_release (assertive.reflection), 24
is_r_release_candidate (assertive.reflection), 24
is_r_revised (assertive.reflection), 24
is_r_slave (assertive.reflection), 24
is_readable_connection (assertive.files), 20
is_readable_file (assertive.files), 19, 129
is_recursive (assertive.properties), 16
is_revo_r (assertive.reflection), 24
is_rstudio (assertive.reflection), 24
is_rstudio_desktop (assertive.reflection), 24
is_rstudio_server (assertive.reflection), 24
is_scalar (assertive), 14
is_scalar (assertive.properties), 14, 16, 21
is_single_character (assertive.strings), 21
is_solaris (assertive.reflection), 24
is_square_matrix (assertive.matrices), 21
is_subset (assertive.sets), 22
is_superset (assertive.sets), 22
is_symmetric_matrix (assertive.matrices), 21
is_true (assertive), 12
is_true (assertive.base), 12, 104
is_true (Rcpp), 12, 104
is_true (testthat), 12, 104
is_unix (assertive.reflection), 24
is_unsorted (assertive.properties), 16
is_upper_triangular_matrix (assertive.matrices), 21
is_valid_variable_name (assertive.code), 27
is_vector (assertive.properties), 16
is_windows (assertive.reflection), 24
is_windows_10 (assertive.reflection), 24
is_writable_file (assertive.files), 19
is_zero_matrix (assertive.matrices), 21
isTRUE (base), 7, 13

knit (knitr), 114

library (base), 8, 19, 22, 23, 27, 28, 36, 49, 50, 58, 66–69, 71, 72, 74, 77–80,
        88, 92, 102, 103, 107, 108, 111, 112, 114, 118, 119, 128–132,
        139–141, 143, 144, 146
lines (graphics), 149
list (base), 14, 63, 80, 109, 144, 147, 149
lm (stats), 23, 49, 93, 100
lm class (stats), 49, 93
lockBinding (base), 27
log (base), 3–5, 29–31, 48
log2 (base), 44, 45, 47, 122
logical vector (base), 119

make_label (testthat), 125, 153
make_label (testthat, internal), 124
Map (base), 149
match.arg (base), 140
matrix (base), 21
max (base), 149
mean (base), 29–32, 49, 65, 72, 123, 133, 138, 139
message (base), 74, 141
min (base), 39–41

na (assertive.base), 141
names (base), 13, 15, 42, 93, 146, 149
natural_log, 5
nchar (base), 131
new.env (base), 27
numeric (base), 14, 39–41, 46–48
numeric vector (base), xiv, 3, 14, 36, 104, 119
nzchar (base), 20, 149

object (roxygen2), 125, 153
on.exit (base), 92, 97
options (base), 63, 91, 92

package_coverage (covr), 88
parent.frame (base), 99
paste (base), 43, 64, 108, 123, 149, 151
paste0 (base), 141, 149
percent_coverage (covr), 88
pnorm (stats), 50
pointsGrob (grid), 68, 69
PostgreSQL (RPostgreSQL), 96, 98
print (base), 10, 40, 67, 68, 70, 71, 121, 136, 146

quantile.default (stats), 73, 142
query (dplyr), 97, 99

R.home (base), 19
Rcpp.package.skeleton (Rcpp), 102
read.csv (utils), 26, 70, 129, 130
read.csv2 (utils), 70
read.dcf (base), 103
read.delim (utils), 70
read.delim2 (utils), 70
read.table (utils), 70, 71
rep_len (base), 4, 5
repeat (base), 44, 45, 47
replicate (base), 51, 137, 138
residuals (stats), 94
return (base), 30, 31, 45, 47, 133, 142, 150
rlnorm (stats), 31, 153
rnorm (stats), 30, 49, 134
round (base), 7
row.names (base), 15, 141
roxygenize (roxygen2), 107, 149
run_testthat_tests (testthat, C routine), 103

sample (base), 31
sapply (base), 48, 137
save (base), 98, 99
saveRDS (base), 39, 95, 108
scalar_with_cause class (assertive.base), 119
search (base), 140
seq_len (base), 4, 5
serializeToConn (base, internal), 109
set_cause (assertive), 121
set_cause (assertive.base), 121, 122, 152
setNames (stats), 143
shQuote (base), 108
sig (sig), 68–71
sig_report (sig), 66, 67
skip_if_not_installed (testthat), 87
skip_on_appveyor (testthat), 87
skip_on_cran (testthat), 87, 109
skip_on_travis (testthat), 87
sprintf (base), 97, 119, 123, 125, 153
sqldf (sqldf), 99
sqrt (base), 35, 37, 43, 44, 49, 81, 135, 137
square_root, 44

square_root_v2, 45, 46
square_root_v3, 47
stdout (base), 107, 108
stop (base), 52
stopifnot (base), 7
str (base), 65
str (utils), 26
str_detect (stringr), 122
stri_detect_regex (stringi), 122
stri_replace_all_regex (stringi), 141
sum (base), 4, 5
summarise_ (dplyr), 139
summarize (dplyr), 72, 139
summarize (plyr), 72
summary (base), 53, 54, 65
suppressPackageStartupMessages (base), 8, 19, 22, 23, 27, 28, 36, 49, 58, 68, 69, 72, 74, 80, 88, 92, 107, 108, 111, 118, 119, 128, 139–141, 144
Sys.Date (base), 50, 132
Sys.getenv (base), 96, 98
Sys.sleep (base), 87
system.file (base), 20

t.test (stats), 95
tapply (base), 138
tempdir (base), 108
test_check (testthat), 51, 78, 95
test_dir (testthat), 51, 53, 78
test_file, 52, 53
test_file (testthat), 51–53
test_that (testthat), 36, 37, 39–52, 81, 91–95, 98–100, 106, 108, 125, 135–138, 141, 146, 149, 151, 153, 154
train.default (caret), 69
trainControl (caret), 69, 140, 141
typeof (base), 48

unlink (base), 80, 144
unlist (base), 149
update_geom_defaults (ggplot2), 63
use_appveyor (devtools), 85
use_catch (devtools), 103
use_catch (testthat), 103, 105, 147
use_coverage (devtools), 89
use_first (assertive), 31
use_first (assertive.base), 133
use_rcpp (devtools), 102, 105, 147

use_testthat (devtools), 79, 81, 145, 147
use_travis (devtools), 85
use_vignette (devtools), 114

vapply (base), 137
vector_with_cause class (assertive.base), 121, 122
Vectorize (base), 137

warning (base), 30, 31, 45, 47, 133
which (base), 149
with (base), 138
with_mock (testthat), 97–100
with_options (withr), 67, 92, 146
write.csv (utils), 71, 108, 141
write.dcf (base), 103
write_ini, 149
writeLines (base), 52, 81, 102, 103, 105, 106, 144, 145, 147, 148

year (data.table), 76, 143